Quantifying the Environment

Stefan Emeis

Measurement Methods in Atmospheric Sciences

In situ and remote

with 103 figures and 28 tables

Gebr. Borntraeger Science Publishers 2010

Author: Stefan Emeis
Institut für Meteorologie und Klimaforschung
Atmosphärische Umweltforschung (IMK-IFU)
Karlsruher Institut für Technologie (KIT)
Kreuzeckbahnstraße 19
82467 Garmisch-Partenkirchen
Germany
stefan.emeis@kit.edu

Cover: Front cover: Radio-acoustic sounding system (SODAR-RASS) remotely observing boundary-layer wind and temperature profiles up to several hundreds of metres above ground (see also Fig. 76). The three white acoustic antennas are seen behind one of the radio antennas in the foreground.
Back cover from left to right: Radio-acoustic sounding system (SODAR-RASS, see also Fig. 76), open-path gas analyzer for CO_2 for turbulent flux measurements (see also Fig. 55), rain gauges with wind shields (see Fig. 34), Stefan Emeis (Photo: Sebastian Emeis).

Figures: All figures and graphs – unless otherwise stated in the legends – are from the author.

ISBN 978-3-443-01066-9

© 2010 Gebrüder Borntraeger, Stuttgart

∞ Printed on permanent paper conforming to ISO 9706-1994

Typesetting: Satzpunkt Ursula Ewert GmbH, Bayreuth
Printed in Germany by Tutte Druckerei GmbH, Salzweg
Publisher: Gebr. Borntraeger Verlagsbuchhandlung, Johannesstr. 3A, 70176 Stuttgart, Germany
www.borntraeger-cramer.de, mail@borntraeger-cramer.de
Information on this title: www.borntraeger-cramer.de/9783443010669

All rights reserved, including translation or to reproduce parts of this book in any form. This book, or parts thereof, may not be reproduced in any form without permission from the publishers.

"I often say that when you measure what you are speaking about and express it in numbers, you know something about it, but when you cannot measure it, when you cannot express it in numbers, then your knowledge is of a meagre and unsatisfactorily kind; it may be the beginning of knowledge, but you have scarcely in your thoughts advanced to the stage of science, whatever the matter may be."

Lord Kelvin (1824–1907)*

The malleefowl (*Leipoa ocellata*), an Australian megapode chicken, builds a nest from twigs and leaves in a up to 5 m wide and 1 m deep hole in the sand and covers it with sand. The bird keeps testing the temperature of the nest by dipping its beak into it. It influences the fermentation of the moist leaves in the nest by adding or removing some sand and thus keeps the temperature within the nest at constantly 33.5 °C.

www.world-of-animals.de/tierlexikon

The snowy tree cricket (*Oecanthus fultoni niveus*) is also called a thermometer cricket because the rate of its chirps is temperature depending. Following Dolbear the temperature in °F can be determined by counting the number of chirps per minute, then subtracting 40, dividing the result by 4, and finally adding 50:

$$t\,[°F] = (n-40)/4 + 50 = (n/4) - 10 + 50 = (n/4) + 40$$

In centigrades the following approximation holds:

$$t\,[°C] = ((n/4) + 40 - 32)\,5/9 = (5n/36) + 40/9 \approx (n/7) + 4.5$$

A.E. Dolbear (1897–1910)**

---

* From: Kelvin, W.T., 1889: Electrical Units of Measurement. In: Kelvin, W.T.: Popular Lectures and Addresses. Vol. I, Macmillan, London, 73.

** From Dolbear, A.E., 1897: The cricket as a thermometer. The American Naturalist, **31**, 970–971.

This publication has been supported by the following companies. All ads have been acquired after the final manuscript has been prepared for publication. Sponsors are listed in alphabetical order.

Campbell Scientific Ltd., Leicestershire, U.K.

GWU-Umwelttechnik GmbH, Erftstadt, Germany

LI-COR Biosciences, Bad Homburg, Germany

METEK GmbH, Elmshorn, Germany

Scintec AG, Rottenburg, Germany

Verein deutscher Ingenieure e.V., Düsseldorf, Germany

Preface

This book has emerged from lectures on meteorological measurement methods at the University of Cologne, Germany, both for students heading for meteorology as a principal subject and for graduated students heading for an International Master on Environmental Sciences (IMES). The teaching for both groups of students required to give a general overview of the large field of different measurement techniques which are available today to determine the state and composition of the Earth's atmosphere. The intention of these lectures was to convey the basic principles of the different observation and monitoring methods. Due to the large number of different measurement principles, it was not possible to describe single methods with a great depth. Additionally, many modern observation methods have become so complex, that a full description would nearly require a special series of lectures of its own.

The intention of this publication is to provide the reader with a manual which informs her or him on the available techniques of measurement, and if several methods are available, to give advices which method to choose for a given task. For this purpose, the lecture notes have been extended in several ways. First of all, more than 400 references, mostly to the recent scientific literature, have been added which will help the reader of this book to get deeper insight in the various methods. Furthermore, tables at the beginning of each major section in Chapters 3 to 7 serve for a quick reference to the different methods. Then, recommendations have been added after each major section which contain additional hints for the use of some instruments and which will facilitate the selection of a proper instrument for a given task. Finally, an appendix has been supplied which gives an impression of an additional body of literature which is often mandatory to be consulted when planning an observation or a monitoring task: national and international guidelines and standards. Today, the work of many bureaus of standards also covers a larger number of environmental monitoring techniques and gives precise information on the application of the instruments and the techniques of data evaluation in order to enable the acquisition of reliable and comparable data sets. A special index will help to use this information.

Many people encouraged and supported me during the compilation of the material for this book and helped with hints to further measurement methods, to special papers in the literature, and to illustrative material. First of all, my thanks go to the publisher who accompanied the preparation of the material with great interest from the very beginning and who offered this fine publication platform to me. Several colleagues from instrument manufactures read parts of the manuscript and gave valuable hints and supplied a few graphs. A great thank goes to Helmut Mayer, who read an earlier version of the complete manuscript, and even more important, who made his large collection of illustrative material available to me. Several of his photographs have been selected and will help the reader to better understand the measurement techniques. Nearly unestimatable is the support which a native English speaker can give an author who has grown up with a language different from English. Here, I experienced the great luck to get the help from Richard Foreman, who diligently read the complete manuscript and who made an uncountable number of suggestions to improve the text.

Now, I hope, the book will be useful to its readers. As everybody knows, no one is perfect and a careful and experienced reader will easily spot those methods described in the book, with which the author is more acquainted than with others. Nevertheless, also with help of the many references included in the text, hopefully a fair and balanced overview on the presently available observation and monitoring techniques has been achieved. May the information gathered in this publication contribute a small piece to the making of a better and sustainable environment for all of us and those following us.

Weilheim, October 2009

Contents

Preface VII
1 Introduction. 1
   1.1   The necessity for measurements 2
   1.2   Definition of a measurement 3
   1.3   Historical aspects................................... 4

2 Measurement basics 7
   2.1   Overview of methods 7
         2.1.1   Direct and indirect methods 7
         2.1.2   In-situ and remote sensing methods 7
         2.1.3   Instantaneous and integrating methods 8
         2.1.4   On-line and off-line methods, post-processing .. 9
         2.1.5   Flux measurements 9
   2.2   Main measurement principles 10
   2.3   Measurements by inversion........................... 12
         2.3.1   Inversion with one variable.................. 12
         2.3.2   Inversion with more than one variable 14
         2.3.3   Well-posed and ill-posed problems 16
   2.4   Measurement instruments 16
         2.4.1   Active and passive instruments............... 16
         2.4.2   Analogue and digital instruments 17
   2.5   Measurement platforms.............................. 18
   2.6   Measurement variables 22
   2.7   General characteristics of measured data 23
   2.8   Data logging 26
   2.9   Quality assurance/quality control 27

3 In-situ measurements of state variables 29
   3.1   Thermometers 29
         3.1.1   Liquid-in-glass thermometers................. 31
         3.1.2   Bimetal thermometers....................... 33
         3.1.3   Resistance thermometers, thermistors.......... 34
         3.1.4   Thermocouples, thermopiles.................. 35
         3.1.5   Sonic thermometry 36
         3.1.6   Measurement of infrared radiation 37
         3.1.7   Soil thermometer........................... 38
         3.1.8   Recommendations for temperature measurements 38
   3.2   Measuring moisture................................. 40
         3.2.1   Hygrometer 43
         3.2.2   Psychrometers............................. 44
         3.2.3   Dewpoint determination 45
         3.2.4   Capacitive methods......................... 46

		3.2.5	Recommendations for humidity measurements	46
	3.3	Pressure sensors.		47
		3.3.1	Barometers	48
		3.3.2	Hypsometers	50
		3.3.3	Electronic barometers	51
		3.3.4	Microbarometer.	52
		3.3.5	Pressure balance	52
		3.3.6	Recommendations for pressure measurements	53
	3.4	Wind measurements		53
		3.4.1	Estimation from visual observations	57
		3.4.2	Wind direction.	57
		3.4.3	Cup anemometer.	58
		3.4.4	Pressure tube	59
		3.4.5	Hot wire anemometer	61
		3.4.6	Ultrasonic anemometer	61
		3.4.7	Propeller anemometer	62
		3.4.8	Recommendations for wind measurements	63
4	In-situ methods for observing liquid water and ice			64
	4.1	Precipitation.		64
		4.1.1	Rain sensors (Present Weather Sensors)	65
		4.1.2	Rain gauges (totalisators)	66
		4.1.3	Pluviographs	67
		4.1.4	Disdrometer.	67
		4.1.5	Special instruments for snow	68
		4.1.6	Recommendations for precipitation measurements	69
	4.2	Soil moisture		70
		4.2.1	Gravimetric methods.	70
		4.2.2	Neutron probes	70
		4.2.3	Time domain reflectrometry (TDR)	70
		4.2.4	Tensiometers	71
		4.2.5	Resistance block tensiometer	71
		4.2.6	Recommendations for soil moisture measurements	72
5	In-situ measurement of trace substances			73
	5.1	Measurement of trace gases.		74
		5.1.1	Physical methods.	76
		5.1.2	Chemical methods.	81
		5.1.3	Recommendations for the measurement of trace gases	84
	5.2	Particle measurements.		84
		5.2.1	Determination of the particle mass	85
		5.2.2	Measuring particle size distributions	88
		5.2.3	Measurement of the chemical composition of particles	92
		5.2.4	Measuring the particle structure	94

		5.2.5	Saltiphon	94
		5.2.6	Recommendations for particle measurements	94
	5.3	Olfactometry		95
	5.4	Radioactivity		96
		5.4.1	Counter tubes	96
		5.4.2	Scintillation counters	97
		5.4.3	Recommendations for radioactivity monitoring	97
6	In-situ flux measurements			98
	6.1	Measuring radiation		98
		6.1.1	Measuring direct solar radiation	100
		6.1.2	Measuring shortwave irradiance	100
		6.1.3	Measuring longwave irradiance	103
		6.1.4	Measuring the total irradiance	103
		6.1.5	Measuring chill	104
		6.1.6	Sunshine recorder	104
		6.1.7	Recommendations for radiation measurements	105
	6.2	Visual range		105
	6.3	Micrometeorological flux measurements		106
		6.3.1	Cuvettes	108
		6.3.2	Surface chambers	108
		6.3.3	Mass balance method	110
		6.3.4	Inferential method	110
		6.3.5	Gradient method	111
		6.3.6	Bowen-ratio method	112
		6.3.7	Flux variance method	112
		6.3.8	Dissipation method	113
		6.3.9	Eddy covariance method	113
		6.3.10	Eddy accumulation methods	117
		6.3.11	Disjunct eddy covariance method	118
		6.3.12	Recommendations for the measurement of turbulent fluxes	118
	6.4	Evaporation		119
		6.4.1	Atmometers	119
		6.4.2	Lysimeters	120
		6.4.3	Evaporation pans and tanks	121
		6.4.4	Recommendations for evaporation measurements	121
	6.5	Soil heat flux		122
	6.6	Inverse emission flux modelling		122
7	Remote sensing methods			124
	7.1	Basics of remote sensing		124
	7.2	Active sounding methods		129
		7.2.1	RADAR	129

		7.2.2	Windprofilers	133
		7.2.3	SODAR	135
		7.2.4	RASS	141
		7.2.5	LIDAR	143
		7.2.6	Further LIDAR techniques	151
	7.3	Active path-averaging methods		152
		7.3.1	Scintillometers	152
		7.3.2	FTIR	153
		7.3.3	DOAS	155
		7.3.4	Quantum cascade laser	156
	7.4	Passive methods		157
		7.4.1	Radiometers	157
		7.4.2	Photometers	159
		7.4.3	Infrared-Interferometer	160
	7.5	Tomography		160
		7.5.1	Simultaneous Iterative Reconstruction Technique (SIRT)	161
		7.5.2	Algebraic Reconstruction Technique (ART)	161
		7.5.3	Smooth Basis Function Minimization (SBFM)	162
8	Remote sensing of atmospheric state variables			163
	8.1	Temperature		163
		8.1.1	Near-surface temperatures	163
		8.1.2	Temperature profiles	164
	8.2	Gaseous humidity		167
		8.2.1	Integral water vapour content	167
		8.2.2	Vertical profiles	167
		8.2.3	Large-scale humidity distribution	168
	8.3	Wind and turbulence		170
		8.3.1	Small-scale near-surface turbulence	170
		8.3.2	Horizontal wind fields	171
		8.3.3	Vertical wind profiles	172
		8.3.4	Turbulence profiles	176
		8.3.5	Cloud winds	176
		8.3.6	Ionospheric winds	176
	8.4	Mixing-layer heights		177
		8.4.1	LIDAR	177
		8.4.2	SODAR	179
	8.5	Turbulent fluxes		180
	8.6	Ionospheric electron densities		181
	8.7	Recommendations for remote sensing of state variables		181
9	Remote sensing of water and ice			184
	9.1	Precipitation		184
		9.1.1	RADAR	184

		9.1.2 Precipitation measurements from satellites.	186
	9.2	Clouds	187
		9.2.1 Cloud base.	187
		9.2.2 Cloud cover.	188
		9.2.3 Cloud movement.	188
		9.2.4 Water content.	189
	9.3	Recommendations for remote sensing of liquid water and ice	189
10	Remote sensing of trace substances		190
	10.1	Trace gases	190
		10.1.1 Horizontal path-averaging methods	191
		10.1.2 Vertical column densities	191
		10.1.3 Sounding methods.	192
	10.2	Aerosols.	193
		10.2.1 Aerosol optical depths (AOD)	194
		10.2.2 Sounding methods.	195
	10.3	Recommendations for remote sensing of trace substances	197
11	Remote sensing of surface properties.		198
	11.1	Properties of the solid surface	199
		11.1.1 Surface roughness	199
		11.1.2 Land surface temperature	199
		11.1.3 Soil moisture	199
		11.1.4 Vegetation	200
		11.1.5 Snow and ice	201
		11.1.6 Fires.	201
	11.2	Properties of the ocean surface	202
		11.2.1 Altitudes of the sea surface	202
		11.2.2 Wave heights	202
		11.2.3 Sea surface temperature	203
		11.2.4 Salinity	203
		11.2.5 Ocean currents.	204
		11.2.6 Ice cover, size of ice floes	204
		11.2.7 Algae and suspended sediment concentrations.	204
12	Remote sensing of electrical phenomena		205
	12.1	Spherics.	205
		12.1.1 Directional analyses	205
		12.1.2 Distance analyses	205
	12.2	Optical lightning detection	206
13	Outlook on new developments		207

Literature	209
Subject index	231
Appendix: Technical guidelines and standards	241
Index to the Appendix	255

1 Introduction

Without measurements we cannot gain any cognition on the exact state of our environment. This relates to the atmosphere as well as to all other compartments of the Earth's system. Taking a measurement is thus one of the most basic and elementary exercises in each science. This very fact in itself justifies a monography on meteorological measurement techniques and methods.

After the basics of air temperature, pressure, and moisture measurements have already been laid three to four hundred years ago, meteorological measurement techniques have seen a rapid development in the last few decades. Fostered by the progress in the development in electronics and computer technology, smaller and more complex measuring devices became available. At the end of the 1950s, in addition, a completely new measurement platform became available: satellites in orbits around the Earth. Based on this evolution an entirely new type of instruments came into existence: remote sensing devices. Today remote sensing techniques comprise a larger part of all meteorological observation techniques. Therefore nearly the same space as for the classical methods is devoted to remote sensing methods in the present book.

With the growing awareness for environmental protection a shift in observational needs has taken place compared with classical meteorology. Air quality and chemistry have become an important issue to observe and monitor. The spectrum of these new requirements ranges from acid rain and near-surface air pollution to stratospheric ozone loss and the still undamped increase of the concentration of greenhouse gases in the Earth's atmosphere. This justifies separate chapters on the measurement of atmospheric trace gases and aerosols to be included in this book.

Since the last two centuries has seen an advancing specialisation in meteorology as in most other scientific disciplines, it now becomes more and more obvious that only a joint consideration of all compartments of the biogeochemical system Earth is meaningful. This includes the observation of exchange processes between these compartments. For example, the growing importance of monitoring exchanges between the atmosphere and its underlying surface has recently increased once again the interest in micrometeorology and its special observational skills. Here, the focus is especially aimed on the direct measurement of turbulent fluxes of carbon dioxide and water vapour. Although monographs on micrometeorology exist, micrometeorological methods must be a part of any general overview on meteorological observational techniques.

A few monographs on observational techniques for the atmosphere are available today which aim at a similar completeness as the classical book by Kleinschmidt (1935), e.g. "Meteorological Measurement Systems" by Brocks & Richardson (2001), Strangeways' "Measuring the Natural Environment" in a second edition from 2003, "Surface Meteorological Instruments and Measurements Practices" by Shrivastava (2008), and the latest edition of the WMO "Guide to Meteorological Instruments and Methods of Observation" (WMO 2008) appeared recently. Strangeways' book predominantly concentrates on classical techniques emphasizing some hydrological aspects; Brocks & Richardson (2001) likewise concentrate on classical meteorological techniques, while the WMO publication quite naturally focuses on the needs of the weather services, as does the publication by Shrivastava.

Therefore the present book tries to offer a more comprehensive review on atmospheric measurement techniques within which air chemistry, remote sensing, and micrometeorology are given the necessary space they require today. However, remote sensing is still a field of rapid development of which no final, ultimate description of all relevant measurement techniques can be provided here. The remote sensing methods introduced here rather represent a snapshot of the present state of development. The book intends to present an overview of the large variety of instruments and their basic operational principles. There is no space to go into technical details of single instruments which may even change from manufacturer to manufacturer. Such details for single classical meteorological instruments for surface observations can be taken from, e.g. WMO (2006) and Shrivastava (2008). A large number of references and the list of guidelines and standards in the Appendix will help the reader to find more details on special instruments and methods.

Data acquisition and storage is a task which is nowadays nearly completely automated. Thus, these two tasks will be given only minor consideration here. This however, should not be interpreted as an invitation to use the available hard- and software for data management without a critical perspective. The emphasis of this book will therefore be on the basic principles behind the measurement techniques. The knowledge on these techniques is a prerequisite for reliable interpretation of data.

1.1 The necessity for measurements

Observation is a cornerstone for gaining cognition and understanding in all sciences. Measurements quantify observations. Quantifying data acquisition has been the precondition for all modern natural sciences and thus also for physics and meteorology. Knowledge on past and present processes taking place in the atmosphere can only be obtained by observations and measurements. Likewise, measurements lay the foundation for any forecast of future events, and only measurements will permit later tests of the accuracy and precision of these forecasts. Just as well, measurement techniques and methods are the prerequisite for experimental studies in order to improve our understanding of selected processes.

In applied meteorology, measurements are necessary for analysis, monitoring, and documentation of the present weather and climate, as well as for the acquisition of initial and input data for models of weather and climate predictions. Likewise, they are needed for assessment and for monitoring air quality and to check for complience with the given threshold and limit values. In meteorological research, measurements are mandatory for any further enhancement of the knowledge. All these tasks demand high-quality measurement devices. More rigorous research and monitoring demands are the main driving force for the ongoing development and refinement of measurement techniques and instrumentation.

1.2   Definition of a measurement

Measurements aim at a quantitative determination of the state of a system (in meteorology this system is usually the atmosphere or a part of it) with respect to selected thermodynamic, kinetic, and chemical state variables. For an ideal gas such as dry air e.g. two of the three basic state variables pressure p, volume V, and temperature T must be determined. Then the thermodynamic state of this gas is known because the third variable can be computed from the law for ideal gases ($pV = RT$ with the gas constant R). In order to assess the kinetic state of a gas, its movement has to be determined, i.e. the spatial distribution of the three velocity components must be measured in a gas volume. Its chemical state can be inferred from an analysis of the chemical composition of the gas. For a description of the temporal evolution or for a forecast of a system, the knowledge of internal conversion rates and of energy and mass fluxes across the outer boundaries of the system have to be known. These include, e.g., the incoming solar radiation, the outgoing thermal radiation, mass fluxes such as falling precipitation, or turbulent vertical fluxes at the Earth's surface.

To be very precise, a measurement is performed by comparing a selected state variable of the system in question (pressure, temperature, etc.) with a predefined scale for this variable (Pascal for pressure, Kelvin for temperature, etc.). This comparison yields a number indicating how often the selected scale is contained in the measured quantity. Or putting it differently, this number establishes a relationship between the state of the observed system and a selected system of reference which serves as a scale.

Practically, a measurement is performed by operating a measuring device. These devices capture the physical or chemical state of the observed system in a suitable way and perform (usually after a gauging or calibration procedure) the above mentioned comparison. To define what is best suited will be the main task of the present book.

The recorded number or measurement value obtained from a measurement cannot be judged independently of that particular measurement technique or method. Peculiarities and limitations of the technique must be known. These comprise, e.g., the temporal and spatial resolution of the data measured with a given instrument. Variations of the measured value smaller than the resolution of the instrument cannot be assessed. Or to put it in another way: a fishing net with a 5 cm mesh size will catch only fish larger than 5 cm (Dürr 1988). Principally, a measuring instrument will only deliver information on processes for which it has been designed.

Measurement of all relevant state parameters of a system results in a set of numbers (parameters) which characterizes and parameterizes the system. This does not necessarily imply a complete description of this system because a choice of what is relevant has been determined or – using the net analogy introduced above – a fishing net with a predefined mesh size has been employed.

1.3   Historical aspects

Quantifying instruments have been known for about 400 years. For information from the time before the invention of modern instruments, one had to rely on "proxy data". These are observations of facts and circumstances which are related to the sought after variable in some, mostly not very precise way. Proxy data include, e.g., the width of tree rings, carbon and oxygen isotope ratios in organic matter, data on harvesting and floods. We will not address the methods to survey and evaluate proxy data in the present book.

The still ongoing development of instruments, that deliver quantitative and reproducible data, started in the period of the renaissance in the 17th century. During the past 30 years, semiconductors and microelectronics have been offering entirely new perspectives. Today, progress in computer technology and storage media permit nearly completely automated measurements, the storage of vast amounts of data, and online evaluations. The precision and resolution of many observational methods has been enhanced considerably – and at the same time – the efforts for maintenance reduced drastically.

Today, in addition to the classical methods which require the direct proximity of the instrument to the object of investigation, remote sensing techniques gain an ever larger part of the usual measurement techniques. The development of remote sensing instruments has been the prerequisite for the use of new platforms like orbiting satellites. This development has only been possible because mathematical methods such as Fast Fourier Transforms furnished the necessary evaluation techniques. These rapidly operating mathematical tools are necessary because of the indirect nature of remote sensing methods which just record irradiance data: classical meteorological or air quality variables can only be isolated by laborious inversion techniques from remote sensing data. But also classical measurement methods – as the next chapter will show – nearly always involve inversion procedures. However, these inversions are, in most cases, an integral part of the method, either by the technical design of the instrument or by gauging and calibration procedures, and are therefore often not obvious to the users.

Milestones of measurement instrument development have been among others the following developments (for a detailed overview on the invention of meteorological instruments until the middle of the 20th century see Middleton 1969):

1.3 Historical aspects

Fig. 1. Kite used by the meteorological observatory in Lindenberg, Germany, for meteorological measurements throughout the whole troposphere at the beginning of the 20th century.

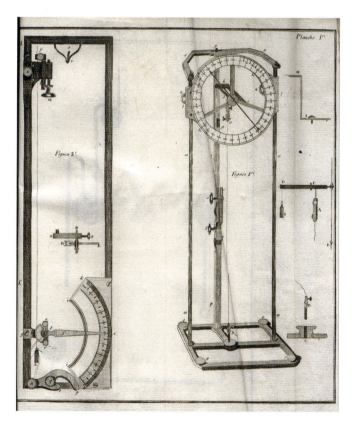

Fig. 2. Drawing showing Saussure's hair hygrometer. Left: the length change of the hair is directly transferred to the pointer (at the lower end of the instrument), right: length change is translated to a clockwise rotating pointer (to the top of the instrument). Source: Saussure (1783), http://orpheus.ucsd.edu/speccoll/weather/b4161877.html (Official Web Page of the University of California San Diego, Copyright 2007, UC Regents).

1593 Galilei's thermoscop (a precursor of the thermometer)
1643 Torricelli's mercury barometer
1654 first meteorological observational network in the Tuscany
1749 first scientific kite ascent (see Fig. 1)
1783 Saussure's hair hygrometer (Saussure 1783, see Fig. 2)
1783 first meteorological measurements during a balloon ascent (first vertical profiling)
1843 Vidie's aneroid barometer
1846 Robinson's wind speed meter
1853 Campell's prototype of the sunshine autographe
1886 first alpine mountain observatory (Hoher Sonnblick, Austria)
1921 first ozone column density measurement with a spectrometer (first remote sensing)
1930 first radiosondes (first regular vertical soundings)
1947 first weather RADAR (first regular detection of precipitation)
1960 first meteorological satellite (first sounding of the Earth's atmosphere from space)
1975 first wind profiler (advancement of ground-based remote sensing)
1979 first DOAS (advancement of optical remote sensing)

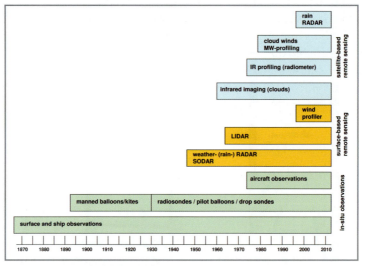

Fig. 3. Timeline of routine weather observation techniques (extended from Uppala et al. 2005) showing in-situ techniques in the lower three bars, surface-based remote sensing techniques in the middle, and satellite-based techniques in the upper four bars.

Figure 3 shows in schematic form the temporal development of routine meteorological measurement methods for monitoring and the provision of input data for numerical weather forecasts.

2 Measurement basics

2.1 Overview of methods

This chapter introduces the basic characteristics of measurement techniques and methods described in this book.

2.1.1 Direct and indirect methods

A direct measurement consists of a direct comparison of the object which is to be studied with a given scale. This simple approach is applicable only for length, mass, and time measurements. A length measurement is accomplished by holding a meter stick directly to the object in question. A time measurement is performed by direct comparison with a clock. A mass determination is made by applying the correct counterweight to the other side of a pair of balances.

All other measurements are carried out indirectly and are based on the observation of how the measurement variable of interest influences or modifies an appropriate sensor. Watching the modifications of the sensor gives indirect access to the variation of the variable of interest. E.g. a radiation measurement is usually based on the observation of how a black plate, which is used as a sensor, heats up or cools down. In this sense, nearly all methods covered in this book are indirect methods. The only exceptions are the three direct methods described above. Mathematically speaking, indirect methods involve an inversion procedure.

2.1.2 In-situ and remote sensing methods

During an in-situ measurement, direct contact prevails between the sensor and the object whose properties are to be determined. This in turn has the consequence that every in-situ measurement influences the object during the measurement. One of the challenges in instrument development is to design it so that repercussions of the instrument on the object are as small as possible. Measuring the air temperature with a thermometer which initially has a temperature different from that of the air will inevitably slightly change the air temperature. Thus the measured temperature is no longer the 'true' air temperature just before the measurement. This repercussion is proportional to the mass ratio of the sensor to the air volume whose temperature is to be determined. Therefore minimizing the size of a sensor is usually a good means to reduce repercussions of sensors on the observed systems.

During remote sensing measurements there is no direct contact between the instrument and the object and repercussions are unlikely. Remote sensing analyses the radiation scattered back or emitted from an object or transmitted through that object. The measurement can be complicated by further modifications of the radiation between the observed system and the instrument. Generally, a radiative transfer equation has to be solved in order to complete the measurement. Passive remote sensing, a method that passively monitors the radiation emitted from a system or transmitted through it, does not influence the observed system at all. Active remote sensing, which is based on the reception of backscattered radiation of a well-defined signal that had been emitted before by the measurement device, could have slight but usually unimportant repercussions on the observed system. The great advantage of remote sensing is that it can aquire information on systems not accessible for in-situ measurements. This comprises, e.g., systems that are unreachable because they are too high above the ground or systems whose state does not permit in-situ measurement (e.g., because they are too hot).

Remote sensing can be performed in several modes: the most important modes are sensing, sounding, scanning, and imaging (see also Fig. 63). Sensing just monitors the instantaneously oncoming radiation. Sounding is based on the emittance of a pulse and the subsequent detection of backscattered radiation during a given time window. Scanning means that the receiving detector is repeatedly turned into different directions in order to detect the incoming radiation. Finally, imaging involves the operation of a focussing element which projects the incoming radiation on a horizontally resolving detector array. Classical photography was the first optical imaging method. Sounding and scanning can be combined, see, e.g., the weather RADAR below.

2.1.3 Instantaneous and integrating methods

Instantaneous measurements are performed in parallel to real processes and yield a momentary value or a series (e.g., time series). The time resolution depends on the sampling rate and the inertia of the sensor.

Integrating measurements either accumulate the impact on the sensor over a longer time period (e.g., daily precipitation sums) or they average over a larger number of instantaneous measurements (e.g., eddy correlation flux measurements). Averaging over many instantaneous measurements is advisable if single measurements have a large associated statistical uncertainty. The averaging procedure then helps to enhance the signal-to-noise ratio.

2.1.4 On-line and off-line methods, post-processing

Both in-situ and remote sensing measurements are mostly on-line methods, i.e. sampling and analysis are made at the place for which and from which the information is needed. This is true for instantaneous as well as for integrating methods.

Some measurement methods are so laborious that they cannot be performed on-line. Post-processing of the obtained data is either very time consuming or needs special boundary conditions in order to deliver reliable results. In these cases raw data are stored and later processed in the laboratory. An example for such a post-processing is the chromatographic analysis of the detailed chemical composition of air collected in special containers. Often off-line methods permit much lower detection limits, i.e. smaller concentrations can be analysed more reliably.

2.1.5 Flux measurements

In many cases characterization of a system does not only require knowledge of state variables such as temperature, density, or trace gas concentration in the interior of the system but also of fluxes of matter or energy across the boundaries of the system. A flux is defined as the amount of a substance, energy, or – more generally – a property which passes per unit time through a unit area. Within the Earth's system, e.g., these can be fluxes from one compartment (atmosphere, ocean, cryosphere, soil, etc.) of the system to another.

The usual transport velocity of substances (gases or aerosols) or energy and momentum is the wind speed. Exceptions are sedimenting atmospheric constituents like rain drops and non-material fluxes of radiation energy. Let A be a surface area, v_a a velocity perpendicular to this surface A, and e a property of the atmosphere. The momentary flux of this property F_e is given by:

$$F_e = \frac{1}{A} v_a e \qquad (2.1)$$

Now, an overbar ($\overline{\ldots}$) denotes a spatial or time mean value (ensemble average) and a prime (') a deviation from this mean (fluctuation), then the flux F_e can be splitted into a mean flux $\overline{F_e}$ and a turbulent flux F_e':

$$F_e = \overline{F_e} + F_e' = \frac{1}{A} \overline{v_a} \overline{e} + \frac{1}{A} \overline{v_a' e'} \qquad (2.2)$$

The measurement of mean fluxes $\frac{1}{A} \overline{v_a} \overline{e}$ which are coupled to atmospheric mean motions is relatively simple. It just requires the determination of the mean wind speed $\overline{v_a}$ and the mean value of the property $\overline{e}$ integrated over the area A and a subsequent multiplication.

The mean flux $\overline{F_e}$ is not always the most important contribution to the total flux F_e. Especially close to the surface, the turbulent part F_e' can be considerably larger than the mean part due to the imperviousness of the ground surface. Simultaneously, mechanical (shear) and thermal (buoyancy) generation of turbulence near the surface lead to larger fluctuations of the vertical velocity around its vanishing mean value. If also the property *e* whose flux is to be determined exhibits larger fluctuations and if these fluctuations are correlated with vertical wind fluctuations, then we will observe a large turbulent flux without any mean mass flux. Especially at the boundary between the Earth's surface and the atmosphere, these turbulent fluxes play a prominent role in the global mass and energy budgets. Therefore Chapter 6 is devoted to flux measurement methods.

The direct measurement of turbulent fluxes requires the simultaneous measurement of wind and property fluctuations with high temporal resolution. Subsequently, the covariance $\overline{v_a' e'}$ between the two time series of fluctuations has to be computed from the raw data in order to determine the turbulent flux. The measurement of turbulent fluxes is one of the main tasks of micrometeorological methods described in Chapter 6.2 in more detail.

2.2 Main measurement principles

In order to characterize the properties of a measurement method we must understand the main steps executed during the data acquisition. The principle sequence is as follows: A sensor or detector records the input signal. The signal is then transformed and/or amplified within the measurement device. Finally, the transformed output signal is shown on a display (either analogue or digital) and/or written to a storage medium (see Fig. 4). Most instruments need some auxiliary energy supply to perform recording, transformation and display or storing of the information. This energy can be supplied either mechanically (e.g. by a clockwork) or electrically.

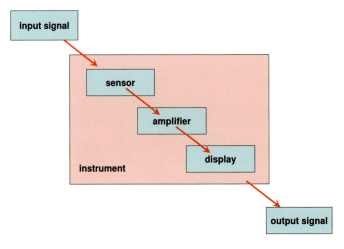

Fig. 4. Basic operation principle of a measurement instrument detecting an input signal and producing an output signal. Instruments usually comprise a sensor, an amplifier, and a display.

2.2 Main measurement principles

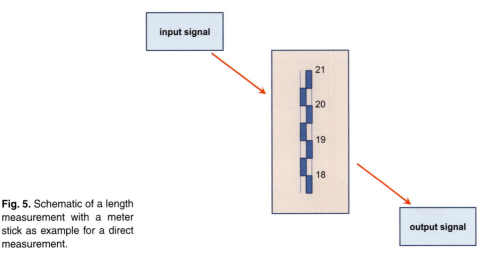

Fig. 5. Schematic of a length measurement with a meter stick as example for a direct measurement.

The simplest example is measuring the length of an object with a meter or yard stick. This stick combines all three functions (sensor, transformer, display) in itself (Fig. 5). The length of the object can directly be read off the scale engraved on the meter stick brought next to the object.

Determining the temperature of an object is somewhat more complicated (Fig. 6). Let us have a look at the classical liquid-in-glass thermometer. The sensor for the air temperature is the thermometric liquid (usually alcohol or mercury) which is captured in the glass container of the thermometer. This glass container acts as an amplifier because the change in length of the liquid thread in the capillary tube depends on the ratio of the width of the capillary tube and the thermometer bulb (see also Fig. 7). The thinner the capillary tube the more sensitive is the thermometer. The display

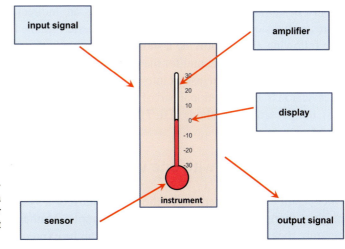

Fig. 6. Schematic of a temperature measurement with a liquid-in-glass thermometer as example for an indirect measurement.

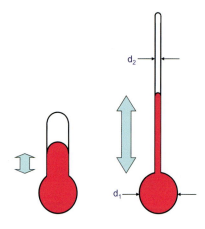

Fig. 7. Mechanical amplification of the temperature signal in a liquid-in-glass thermometer. The amplification is proportional to the ratio d_1 (diameter of the bulb) to d_2 (diameter of the capillary tube). Left: low amplification, right: high amplification.

of the thermometer is the capillary tube to which a scale is attached. Reading the length of the liquid thread in the capillary tube in itself is a direct length measurement as described above.

This makes obvious that the process of obtaining the temperature of a system is an indirect measurement method. The internal energy of the system acts on the thermometric liquid and changes its volume. By the way the thermometer is constructed this volume change of the liquid is translated to a length change of the liquid thread, and finally, this length is read by comparing it to a length scale. Once a thermometer has been built, the scale has to be adjusted to this very thermometer by calibrating it. This calibration process also compensates for effects of the slight change of volume of the glass container of the thermometer with temperature.

2.3 Measurements by inversion

As just discussed, most measurement methods are indirect methods and therefore require an inversion to obtain the sought atmospheric variable from the recorded raw data. The following aims to provide a short introduction to the main idea of inversion. This digression is not fundamental for the understanding of the rest of the book but it is presented here to provide an insight into such formalism. In a much more complex manner, an inversion is a constitutional part of such methods as emission rate determination (Ch. 6.6), remote sensing methods (Ch. 7), and tomographic methods (Ch. 7.5).

2.3.1 Inversion with one variable

Once again the measurement of the temperature of a system with a classical liquid-in-glass thermometer shall serve as an example. Mathematically speaking, a temperature measurement is a mapping which can be expressed by a function g:

2.3 Measurements by inversion

$$m = g(z) \tag{2.3}$$

Function g in (2.3) maps the state variable internal energy, z of the observed system to the measured value, m. Generally, z and m are vectors, but in our example they are just scalar quantities. If the state variable z and the function g are known, the measured value m can be predicted. This is also called forward modelling. Taking a measurement is just the reverse operation: here the measured value m is determined by reading the instrument and we must invert the process described by the function g in order to get the atmospheric state variable z. Mathematically this reads:

$$z = g^{-1}(m) \tag{2.4}$$

As we execute the model described by the function g in an inverse way, the procedure (2.4) is also called inverse modelling. To elucidate this inversion procedure we will specify the functions g and g^{-1} for the liquid-in-glass thermometer. As discussed in Chapter 2.2, measuring the temperature T is a two-step procedure, and therefore we must split the model described by the function g in two parts g_1 and g_2. g_1 mirrors the effect of the internal energy of the air $z = c_v T$ on the volume, v of the thermometric liquid and g_2 transforms the volume change of the liquid, $v(T) - v(T_0)$ to the length change of the liquid thread, $l - l_0$ in the capillary thread:

$$g_1: v(T) = v(T_0)(1 + \gamma(T - T_0)) \tag{2.5}$$

with the thermal expansion coefficient, γ of the thermometric liquid and the temperature, T in K, and:

$$g_2: l = l_0 + \lambda(v(T) - v(T_0)) \tag{2.6}$$

with the constant, λ which has the dimension m^{-2} and which depends only on the construction of the glass container of the thermometer. l_0 is the length of the liquid thread at temperature T_0.

The length of the liquid thread is measured directly (expressed by a function d) by reading the scale fixed at the thermometer. The complete concatenated forward model for a temperature measurement reads:

$$m = d(l) = d(g_2(v)) = d(g_2(g_1(T))) \tag{2.7}$$

Inserting (2.5) and (2.6) in (2.7) we obtain for (2.3):

$$m = d(l_0 + \lambda(v(T_0)(1 + \gamma(T - T_0)) - v(T_0))) = d(l_0 + \lambda\gamma v(T_0)(T - T_0)). \tag{2.8}$$

As we need (2.4) instead of (2.3) we now have to perform an inversion. The inverse of the direct observation d^{-1} is identical to d itself. For functions g_1 and g_2 it has to be assumed that they are uniquely invertible. The volume change reacts to a change of the temperature:

$$g_1^{-1}: T = T_0 + (v(T) - v(T_0)) / \gamma v(T_0) \tag{2.9}$$

and the length change reacts to a volume change:

$$g_2^{-1}: v(T) = v(T_0) + (l - l_0) / \lambda, \tag{2.10}$$

where (2.9) is thus the inverse of (2.5) and (2.10) the inverse of (2.6). We now can specify the inverse model to the forward model (2.7) as:

$$T = g_1^{-1}(v) = g_1^{-1}(g_2^{-1}(l)) = g_1^{-1}(g_2^{-1}(d(m))) \tag{2.11}$$

Substituting (2.9) and (2.10) in (2.11) finally yields the wanted inverse model of a temperature measurement:

$$T = T_0 + ((v(T_0) + (d(m) - l_0) / \lambda) - v(T_0)) / \gamma v(T_0) = T_0 + (d(m) - l_0) / (\lambda \gamma v(T_0)) \tag{2.12}$$

The inverse model (2.12) permits deriving the temperature T from a reading m of the instrument, because γ and λ are known due to our choice of the thermometric liquid and the construction of the instrument, and l_0 and T_0 are obtained through a calibration of the thermometer.

2.3.2 Inversion with more than one variable

Frequently, specifying the inverse model (2.12) is much more complex. The reading of a mercury barometer may serve as an example. The observed length of the mercury thread in the tube of the barometer does not only depend on the ambient air pressure but also on the temperature of the barometer itself. For this reason classical mercury barometers always have an integrated thermometer to determine the temperature of the instrument. This temperature influences both the volume of the barometer tube and the volume of the mercury within the barometer. In Chapter 2.3.1 the air pressure has not been considered because it had been assumed that the glass container of the thermometer is not deformed by the air pressure and therefore the air pressure does not exert any force on the thermometric liquid inside the thermometer. Mathematically, the measurement of the ambient air pressure is described by the three functions g (length change of the mercury pile in the tube of the barometer due to air pressure and air temperature changes), d_1 (reading of the height of the mercury pile in the tube), and d_2 (reading of the thermometer attached to the barometer):

2.3 Measurements by inversion

$$g: \quad l(p,T) = l_p(p_0) + (p - p_0)/(\rho(p, T)\, g) \qquad (2.13)$$

$$d_1: \quad m_p = d_1(l_p) \qquad (2.14)$$

$$d_2: \quad m_t = d_2(l_t) \qquad (2.15)$$

Here p and T are air pressure and temperature of the barometer, ρ is the density of mercury, and g the gravitational force of the Earth. The readings of the barometer and the thermometer can be expressed as:

$$m_p = d_1(g_1(p,T)) = d(l_p(p_0) + (p - p_0)/\rho(p,T)\, g) \qquad (2.16)$$

$$m_t = d_2(l_{t,0} + \lambda \gamma v(T_0)(T - T_0)) \qquad (2.17)$$

with the read pressure value m_p from the length of the mercury pile in the tube l_p, and the read temperature of the barometer m_t from the length of the liquid thread of the thermometer l_t. The state variables, pressure p and temperature T, can be summarized into vector Z, the measured values m_p and m_t into vector M, and the mathematical functions involved g_1, d_1, and d_2 into matrix G. The following forward relation holds:

$$M = G\,Z \qquad (2.18)$$

and in order to obtain the vector Z (2.18) must be inverted:

$$Z = G^{-1}\,M. \qquad (2.19)$$

For practical purposes, (2.19) is solved in two steps. First the pressure value is read from the barometer and subsequently a temperature correction is applied to this pressure value. Up to here we have assumed that measurements and readings have been made without any error. Considering a measurement error F (2.18) changes into:

$$M = G\,Z + F \qquad (2.20)$$

which must be inverted. If Z is non-linear and if a first guess Z_a is available for Z we can linearize around the first guess:

$$M = G\,Z_a + K\,(Z - Z_a) + F \qquad (2.21)$$

here $K_{ij} = \partial G_i / \partial Z_j$ is the Jacobian. Several methods are available for performing the inversion. Overviews of these methods can be found e.g. in Westwater (1997), Bennett (2002), Enting (2002) or Cimini et al. (2006).

2.3.3 Well-posed and ill-posed problems

In order to assess possible difficulties which could arise when performing the just introduced inversion tasks, it has to be distinguished whether we deal with a well-posed or an ill-posed problem. In well-posed problems the results are insensitive to the input data and in ill-posed problems the results are very sensitive to the input data. When trying to solve the latter problems, small variations in the input data can lead to large changes in the results. The inversion of a well-posed forward problem leads to an ill-posed task, the inversion of an ill-posed forward problem is usually simple.

This can be illustrated by considering a mercury thermometer as an example once again (Fig. 7). If we need to determine the volume of a certain amount of mercury at a given temperature using (2.5), then this problem is well-posed. The volume expansion coefficient of mercury is 0.000181 K^{-1}, i.e., a variation in temperature of about 55 K leads to a variation of the volume of the mercury within the thermometer of 1 %. Thus the volume of this amount of mercury can easily be determined with high precision. For the inversion task (2.9) this has the consequence that if the volume of the mercury is known with 1 % accuracy then the temperature can vary by 55 K (this corresponds to a 20 % error in the absolute temperature). Thus, determining the temperature from the determined volume of a known amount of mercury is a very ill-posed problem.

This ill-posed problem has been solved by the inventors of the thermometer by the way they designed it. The largest part of the mercury resides in the bulb and only a very small part resides in the very thin capillary tube of the thermometer. We have seen in Chapter 2.2 that this design acts as a very effective amplifier for the measurement values. If the ratio between the diameter of the bulb and the capillary tube is 100:1, then the amplification is a factor of 100. This implies that if the length of the liquid thread in the capillary tube is known with an accuracy of 1 % (related to the full length of the thread) the temperature can be determined with an accuracy of 0.2 % or 0.55 K. In the case of the thermometer the inverse problem turns out to be a well-posed problem. In contrast to this, the forward problem determining the length of the liquid thread from a given temperature is an ill-posed problem. An error of 1 % in the temperature leads to close to 6 % error in the length of the thread.

2.4 Measurement instruments

2.4.1 Active and passive instruments

Passive instruments such as liquid-in-glass thermometers or barometers take the energy of the output signal solely from the input signal. Active instruments have their own mechanical or electric energy supply. They do not take much energy from the input signal. An example for a mechanical energy source is the clockwork which drives the ventilator in a aspirated psychrometer. An example for an electric energy source is the electricity which is necessary to support the electric current in a resistance thermometer.

The following compilation lists the characteristics of passive and active instruments.

a) passive measurement characteristics
- direct contact without transmitting mechanical forces: the sensor touches the medium whose state is to be analyzed (e.g. thermometer, here internal energy is transmitted from the medium towards the sensor)
- direct contact with transmitting mechanical forces: e.g. barometer, cup anemometer, pressure tube, wind vane
- absorption of a trace compound: e.g. water vapour by a hair hygrometer
- accumulation of sedimenting substances (precipitation, dust) in appropriate containers without forced ventilation
- remote sensing: measurement of irradiance with a radiometer

b) active measurement characteristics
- measurement of the boiling temperature of water (determination of air pressure with a hypsometer)
- measurement of the temperature difference between a dry-bulb and a wet-bulb thermometer with forced ventilation (aspirated psychrometer)
- electrical resistance measurement (resistance thermometer)
- chemical reactions (measurement instruments for trace gases)
- active probing (cascade impactor, deposition of aerosols on filter papers) with forced ventilation
- detection of the backscatter of an emitted pulse (SODAR, ...)
- absorption of an emitted pulse (DOAS, DIAL, ...)
- measurement of the propagation speed of an emitted pulse (RASS, sonic anemometer, acoustic tomography)

2.4.2 Analogue and digital instruments

Analogue instruments offer a continuous output signal that depends on the input signal in a well-defined manner. The magnitude of the output signal is usually read from an attached scale.

Digital instruments offer a discrete output signal that is related to certain threshold values of the input signal. The output value is a number. Digital instruments are often operated electrically or electronically, but mechanical digital instruments also exist. An example for the latter is the tipping-bucket rain gauge (see also Fig. 35 below) that counts the number of position changes of the tipping bucket. The amount of rain is later computed by multiplying this number with the amount of rain necessary to provoke one position change of the tipping bucket.

2.5 Measurement platforms

Measurement platforms are fixed or moving objects on which measurement instruments are mounted. Platforms are among others the Earth's surface, thermometer screens (usually 2 m above ground), vehicles (to obtain point measurements from different sites or horizontal profiles), weather ships (there used to be nine weather ships in the North Atlantic in former times), buoys, masts, towers, tethered balloons, radiosondes, aircraft, and satellites. Measurements made from masts, towers, tethered balloons, and radiosondes are usually in-situ measurements; measurements from the surface and from aircraft can be either in-situ or remote sensing measurements. Measurements from satellites are always remote sensing measurements.

In-situ surface and mast measurements deliver data from the surface layer. Tower measurements permit studying the lower 200 to 300 m of the boundary layer (Fig. 12). With tethered balloons soundings up to about 500 m above ground can be performed. But they take some time to raise and lower with a winch. The balloon itself is influencing the flow and a permit from the air traffic authorities to operate them is required. A compromise between a tethered balloon and an aircraft is using a helipod, a container attached to a 15 m long wire underneath a helicopter, whose operation does not require a permit from the authorities. When the helicopter is cruising, the helipod is ahead of the downwash from the rotor of the helicopter. The helipod is suited to record mean and turbulent fluxes (Bange & Roth 1999). Radiosondes deliver single vertical profiles across the complete troposphere and lower stratosphere. Vertical resolution of radiosonde measurements is limited by the speed of ascent of the sondes. Horizontal distributions of mean and turbulent quantities in the free atmosphere can be investigated with aircraft; the spatial resolution is limited by the cruising speed of the aircraft. Alternatively, ultralight aircraft may be used, which are highly manoeuvrable and which can fly at very low speed (Junkermann 2001). The payload of such an ultralight however is very limited. Recording a vertical profile with an aircraft takes several minutes and due to safety reasons the lowest flight level is on the order of 100 m. Some of these platforms will be described in more detail below.

Fig. 8. Thermometer screen of the German Weather Service at the Hohenpeißenberg observatory.

Thermometer screens

A thermometer screen (Stevenson screen) (Fig. 8) is a containment – usually mounted 2 m above the ground – which allows for passive ventilation by louvered sides, back and front. It is painted white and is designed to protect the thermometers and hygrometers they house from radiative heating and precipitation. The doors of the screen should open towards the North in northern hemisphere settings and towards South in the southern hemisphere in order to prevent the sun shining on the instruments during the reading.

Tethered balloons

Tethered balloons (Fig. 9) consist of a streamlined lifting hull filled with hydrogen or helium gas, attached to a wire or tether. The length of the wire is controlled by a winch. About 10 m below the balloon a small vessel contains instruments for in-situ measurement of temperature, moisture, air pressure, and wind. The weight of the payload is quite limited by the lift force of the balloon (several hundred grams to a few kilograms). The data are transmitted to a data logger on the ground by a by radio transmitter. Tethered balloons can be operated at heights of up to several hundred metres as long as wind speeds are not higher than roughly 10 m/s. One sequence of raising and lowering the balloon takes about half an hour. Because the tether which holds the balloon usually bends down due to its own weight, the height of the balloon cannot be determined from the length of the tether, but from the air pressure measurements.

Fig. 9. Tethered balloon with cup anemometer (Photo: Helmut Mayer).

Radiosondes (aerological measurements)

A radiosonde (Fig. 10) is made up of a 4 m³ rubber balloon filled with hydrogen gas and an instrumented gondola mounted more than 30 m below the balloon. The balloon itself weighs about 800 g, the styrofoam gondola carrying the instruments about 650 g. The rate of ascent is about 300 m per minute. The gondola contains a bimetal thermometer, a hair or lithium chloride hygrometer, and an aneroid barometer. Today, the path of the sonde, from which the wind speed is computed, is monitored by GPS. In earlier times theodolites and RADAR were used for this purpose (Phillips et al. 1980). Acquired data are radioed to the ground. When the maximal height of 20 to 30 km is reached the balloon bursts and the gondola returns to the ground on a small parachute. Found instrument gondolas can be sent back to the weather services, some of them pay a little reward for this. Usually they can be restored to working conditions.

Fig. 10. Radiosonde with instruments a few seconds after launching.

Balloons without instruments which just carry a reflector for radio waves are called pilot balloons and are used to determine vertical wind profiles. The word "radiosonde" refers to the radio communication by which the data are transferred to the ground. Sometimes the term "rawinsonde" (**ra**dar **win**d) is used when the additional determination of the wind by tracking the sonde is emphazised.

Constant-level balloons
Constant-level-balloons are balloons of well defined buoyancy that fly at a pre-specified level of given air density. They do not carry any instruments but serve for the detection and visualization of dispersion

Aircraft
Depending on the scale of the phenomenon and the height of the atmospheric layer which is to be studied, different types of aircraft can be employed. Micro-scale structures of the surface layer and the lower part of the boundary layer can be analyzed with remotely controlled small unmanned aircraft (Egger et al. 2002, Spiess et al. 2007, Reuder et al. 2009), small-scale features of the whole boundary layer and the lower troposphere are obtainable from ultralight aircraft (Fig. 11, Junkermann 2001), and large-scale data from the troposphere and lower stratosphere are yielded by jet aircraft. The spatial resolution of the obtained data depends on the cruising speed. The sensors have to be mounted to the outside of the aircraft, e.g. on a boom fixed to the front tip of the aircraft. Data measured from such an upwind boom are not so severely influenced by the presence of the aircraft itself. For a long time, accurate geo-referencing of aircraft data was a big issue because speed measurements aboard an aircraft only yield the true air speeds, i.e. the speed with respect to the surrounding air, which is usually in motion itself. With the introduction of GPS receivers, with which the speed relative to the ground can be computed, this problem was solved. Nowadays many civil aircraft are part of the Aircraft Meteorological Data Relay (AMDAR) measurement network and routinely supply a wealth of data to weather services worldwide. Since August 1994, five Lufthansa A340 aircraft deliver temperature, water vapour, and ozone data within the MOZAIC project (Marenco et al. 1998) whenever they are in the air.

Fig. 11. Take off of an ultralight aircraft equipped with instruments for meteorological and air quality parameters. Such aircraft can reach heights of up to 4000 m above ground. (Photo: Stefan Metzger).

2.5 Measurement platforms

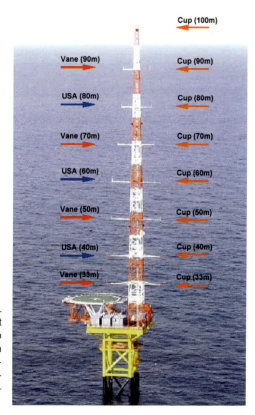

Fig. 12. Offshore meteorological tower FINO1. This 100 m high tower 45 km off the German coast is equipped with cup anemometers from 30 m to 100 m every 10 m and with sonic anemometers in 40, 60, and 80 m height on the tips of the horizontal booms. For some heights wind direction, humidity and pressure are also available. (Photo: Thomas Neumann, DEWI)

Satellites

At the end of the 1950s, the observation of the Earth and its atmosphere from orbiting satellites became possible. Advantages of satellite observations are a large spatial coverage and the opportunity to re-visit selected places of the Earth surface after one or several orbits. In-situ atmospheric observations from satellites are possible for the exosphere only. All other layers of the atmosphere and the Earth surface can only be probed by remote sensing techniques. Available methods with different measurement geometries comprise nadir soundings directly vertical to the Earth surface underneath the satellite, scan soundings at some angle with respect to the vertical, and limb soundings. Passive nadir soundings usually have an insufficient vertical resolution and it is difficult to distinguish between signals originating from the ground and from the various atmospheric layers. Scan soundings usually have quite good horizontal resolution but inherently have the same problems as the direct nadir soundings. Limb soundings offer a better vertical resolution but they take averages over large horizontal areas. They can be either made as emission or absorption measurements with a vertical resolution of roughly 2 km. Emission measurements record radiation emitted from the atmospheric layers without a disturbing signal from the Earth surface. This allows for lower detection limits than for nadir soundings. In absorption measurements (also called occultation measurements) the transmission of light and radiation from natural objects like sun, moon, or bright stars or from artificial objects like GPS satellites through the different atmospheric layers is observed.

A variety of different orbital geometries are used. Normally orbiting satellites are flying from West to East at altitudes of 200 to 850 km above ground and need about 90 to 103 min for one orbit. The orbits are inclined up to about 50° with respect to the Earth's equatorial plane. For regular monitoring purposes, two other orbit geometries are much more important. Polar orbiting satellites fly on orbits inclined 98° with respect to the equator and approximately cross both poles at an altitude of about 850 km. At this altitude, 103 min are required for one orbit and the satellite can perform 14 revolutions around the Earth per day. The orbital plane of polar orbiting satellites is constant with respect to the sun by utilizing the precession effect due to the non-perfect spherical shape of the Earth (Hase & Fischer 2005). This means that these satellites can monitor a swath roughly 2800 km wide of the Earth surface always at the same time of day. Geostationary satellites cruise at an altitude of 36 000 km over the equator. At this altitude their angular velocity is exactly equal to the Earth's, i.e. geostationary satellites seem to be fixed with respect to the Earth surface. These satellites, among them the European METEOSAT satellites, always monitor the same part of the Earth surface (about 40 % of the total surface). Observations of the polar regions from these satellites have low spatial resolution because the satellite sees these regions at a very low angle. The poles themselves cannot be observed from geostationary satellites.

2.6 Measurement variables

The physical and chemical state of any system can be determined from the up to seven basic variables listed in Table 1 using the SI-system (Système international). If applicable, further variables may be derived from these basic variables and are listed in Table 2. In former times (and in some regions of the world still today) further non-metric units or units which contain other time units than seconds are used (e.g. foot as a length unit or horse power as a power unit).

Table 1. Basic variables in the SI system.

variable	unit	unit designation
length	metre	m
time	second	s
mass	kilogramme	kg
temperature	Kelvin	K
electric current	Ampère	A
light intensity	Candela	cd
chemical substance	Mol	mol

Table 2. Derived variables from the basic variables in Tab. 1.

variable	basic units	derived unit	unit designation
area	length · length		m²
volume	length · length · length		m³
velocity	length / time		m / s
acceleration	length / (time)²		m / s²
force	mass · acceleration	Newton	N (kg m / s²)
pressure	force / area	Pascal	Pa (kg / (m s²))
energy	force · length	Joule	J (kg m² / s²)
power	energy / time	Watt	W (kg m² / s³)
electric charge	electric current · time	Coulomb	C (A s)
electric tension (voltage)	force · length / charge	Volt	V (kg m² / (A s³))
electric resistance	voltage / electric current	Ohm	Ω (kg m² / (A² s³))
magnetic field	force / (charge · velocity)	Tesla	T (kg / (A s²))

2.7 General characteristics of measured data

A measured variable can only be interpreted reliably if the properties of the measurement site and the instruments used are known. Then mis-interpretations due to special characteristics of the site or due to limitations of the instrument can be avoided. The following list introduces the main characteristics of a measurement variable. The reader is also referred to Strangeways (2000) or to the definitions in the German standard VDI 3786, part 1 (see Appendix).

Representativity
Measurements taken at a site are always representative for some area upstream of the site. The size of this fetch (also called footprint, Schmid 1994), e.g., increases with the wind speed, decreases with increasing turbulence, increases with the measurement height or with the amount of air sucked in for the analysis of air quality. The measurements are most representative for a site if the upstream area and the site itself have equal characteristics.

Homogeneity
Homogeneity is closely related to representativity. An air flow always tends to reach equilibrium with its underlying surface. Therefore, changes in the surface characteristics (e.g., changes in land use, roughness, soil moisture, heat capacity, etc.) lead to the formation of internal boundary layers whose depth is increasing with increasing distance from the place where the change occurred. The depth of such an internal boundary layer is roughly 100 times this distance. Especially, if measurements are performed in more than one height, the fetches for these measurements should have all the same surface characteristics in order to relate the observed vertical profile properly to the surface characteristics.

Span
The measurement span is the range between the smallest and the largest value that the instrument can record. The span of, e.g., a thermometer is limited by the length of the capillary tube and by the choice of the utilized liquid.

Error (Trueness)
An error designates the statistical or systematic deviation of a measurement value from the true value μ. Statistical or random errors lead to measurement values which are sometimes above the true value and sometimes below it. If an average over a longer series of such measurements x_j is taken it is expected that the mean value $\bar{x}$ approaches the true value μ:

$$\mu = \bar{x} = \frac{1}{n}\sum_{j=1}^{n} x_j, \tag{2.22}$$

where $\bar{x}$ is therefore called the expected value. Systematic errors on the other hand lead to measurement values which bias towards one side. Taking an average reduces the scatter of the measurement values without the mean value converging to the true value. There remains a bias between the mean value and the true value. Systematic errors often originate from insufficiencies or limitations with the utilized instruments or the chosen method. A careful calibration of an instrument before, during, and after the measurement period can considerably reduce the bias (see also the standard DIN ISO 5725-1:1997-11).

Accuracy (Precision)
The accuracy or precision is the inverse of the statistical error. An estimation for the precision is the standard deviation or the root mean square error, determined from n similar measurements, x_j with an expected value, $\bar{x}$:

$$s = \sqrt{\frac{1}{n-1}\sum_{j=1}^{n}(x_j - \bar{x})^2} \tag{2.23}$$

If the error is normally distributed, then about 2/3 of all measurement values fall between $\bar{x} - s$ and $\bar{x} + s$. The result of the measurement is usually written as: $\bar{x} \pm s$ (Kreyszig 1972). A given accuracy for, e.g., a thermometer could read +/– 0.3 °C. Related to the term accuracy is the term repeatability or comparability of a measurement value. The VDI guideline 3786, part 1 (see Appendix), further distinguishes between the 'comparison precision' and the 'repetition precision'. The first can be obtained under comparable ambient conditions, the latter one under identical ambient conditions.

Resolution
The resolution denotes the smallest interval between two different measured values that can be distinguished with a statistical confidence of 95 % or more. The term resolution is not to be mixed up with the term accuracy. Quite frequently the resolution of an instrument can be higher than the accuracy, i.e. it is possible to distinguish

between differences smaller than the accuracy of the instrument. The temporal resolution indicates how small the time difference between two measured values can be so that they still can be considered as independent. The temporal resolution is the inverse measure of the time constant (see below). For sounding or scanning remote sensing devices, the spatial resolution of these instruments must also be given.

Hysteresis
For many measurement instruments it makes a difference whether the measured variable increases or decreases within the span of the instrument. The reason may be the time constant (see below) of the instrument or some specific mechanical, electronic, or chemical properties of the measurement method or the instrument design.

Drift
Many instruments exhibit 'aging' features. I.e., the measurement error increases with time. These can be either due to mechanical or chemical aging processes of the utilized materials and substances. Regular maintenance and calibrations can limit this error.

Time constant (Response time)
The time constant of an instrument is the time which passes until the instrument, after a sudden change of the value that is to be measured, has responded to more than 63 % (= $(1-1/e) \cdot 100$ %) to the new value. The time constant of a thermometer e.g. depends on the specific heat of the thermometer liquid. Closely related are such terms as 'rise time' and 'fall time'.

Detection limit
The detection limit is the smallest value which can be distinguished from zero with confidence (mostly 95 %, Möller 2003). If the measured value is at the detection limit or even below, then the instrument only delivers qualitative but not quantitative information.

Determination limit
The determination limit is a measured value that is with a security of 95 % or more above the detection limit. Following the standard VDI 2449, part 1 (see Appendix), quantitative measurements start at this value (Möller 2003).

Signal-to noise-ratio
The signal-to-noise-ratio (frequently abbreviated as SNR) is the ratio between the measured value and the background noise. A reliable measurement is possible only if the measured quantity clearly peaks from the background noise; and further post-processing of this measured value is advisable.

Sensitivity
The sensitivity is the slope of the calibration line, if a linear dependency exists between the output value of an instrument and the variable to be measured. Sensitivity describes the ability of an instrument to capture small changes of a quantity. Electronic amplification is useless if the background noise is amplified in the same manner as the signal (Möller 2003).

Interference (Selectivity)
Interference or selectivity indicates which amount of a third compound or quantity can erroneously provoke a false measurement signal (Möller 2003).

Specificity
For gas measurements specificity indicates how specific a chosen measurement technique is for a chosen quantity. Specificity is the inverse measure to selectivity (Colls 2002).

Limiting condition of operation
The term 'limiting condition of operation' defines the ambient conditions under which the instrument or the method is able to perform measurements within 95 % of the specified characteristics of the instrument or method.

Availability
Data availability signifies the relative part of the measurement time for which reliable data can be obtained.

2.8 Data logging

In former times, data from meteorological instruments for routine observations were read by observers at fixed times during the day, noted in a logbook, and transmitted

Fig. 13. Electro-mechanical recording of measured data on a paper strip (here: wind direction (left) and wind speed (right)).

telegraphically. Data from recording instruments such as thermo-, hygro-, or pluviographs (Fig. 13) were recorded on stripes of paper, fixed on a turning drum driven by a clockwork, and later analyzed manually. Other devices, such as the sunshine autograph does not require any mechanical driving at all (see Ch. 6.1.6).

Today, digital data loggers are standard. The measured value is usually an electric current or voltage (e.g., a resistance thermometer), or a discrete count (e.g., the number of turns of a cup anemometer). If the instrument delivers an analogue value the transformation of this signal to a digital value must be included in a calibrating or gauging procedure. Data loggers file the measurement results at a specified sampling rate (e.g. 10 min for mean values or 0.1 s for turbulence values). Often radio transmission of the data is done via a cell phone network or using special radio transmission frequency bands. This permits remotely controlling instruments from the laboratory that conducts the measurements (a more detailed discussion of digital measurement recording is given by Foken (2008)).

2.9 Quality assurance / quality control

Reliable measurements require accompanying measures to control quality and accuracy. These procedures are quality assurance and quality control; sometimes summarized in the abbreviation QA/QC.

Quality assurance starts before the first measurement. It begins with a definition of the spatial and temporal resolution of the desired data. Technical rules for the planning of observations can be found, e.g., in VDI 4280, part 1 (see Appendix). Further the expected ranges of the data and the necessary accuracy have to be defined beforehand. These factors influence the choice of the measurement platform and of the instruments employed. The chosen instruments and data loggers have to be tested and calibrated before the start of a measurement campaign. Similar instruments, which must be deployed to different sites during a campaign, should be compared in a comparison experiment at one representative site before the start of the real campaign. This assures that different results from measurements at different sites can be interpreted as spatial differences or gradients later. The same applies to instruments mounted at different heights on a mast or on a tower. Similar checks should also be performed directly after a measurement campaign in order to identify and document possible drifts in the measurement results and instrument sensitivities. Longer measurement campaigns may require calibrations at regular intervals during the campaign. Some instruments also require regular maintenance, e.g., refilling water in aspirated psychrometers, to guarantee a constant data quality during the entire campaign.

Quality control comprises monitoring the running measurements and the preparation of the subsequent data evaluation. This includes the detection of instrument failure and ongoing data consistency checks. Inconsistent data or wrong data must result in an instrument check, i.e. the quality assurance measures described above must be repeated. An inconsistency check has not only to be applied to single data time series but also requires a cross-check between several related data series, through physical or chemical principles. Two values from two different data sets may individually look correct but their combined appearance can be excluded by physical laws (e.g., the absolute moisture cannot exceed a certain temperature-dependent maximum saturation value). Final data series have to be checked for homogeneity. Mean values and standard deviations should not exhibit unexplainable jumps. Outliers, outside a pre-defined span of some multiples of the standard deviation, must be reviewed and explained manually. If necessary they must be eliminated before the data evaluation. Trends in data series, due to instrument drifts, must be identified and eliminated as well. Some methods for data control are given in Foken (2008).

Measurement data whose origin and quality cannot be checked should not be included in a data evaluation.

VDI/DIN Manual
Air Pollution Prevention
VDI Guidelines and DIN Standards for Air Quality

VDI Manuals are compilations of VDI Guidelines and DIN Standards available in DIN A 4 ring binders, on CD-ROM or by download. They are obtainable singly, by subscription including permanent updates or tailored to your technical profile. They focus for example on:

➤ Environmental meteorology
➤ Remote sensing techniques (LIDAR, SODAR, RADAR, FTIR, DOAS)
➤ Analysis and measurement methods for air pollutants (emission, ambient air, particles, gases, trace substances)
➤ Olfactometry

For more information: www.vdi.de/guidelines

Orders and supply: Beuth Verlag GmbH · Burggrafenstraße 6 · D-10787 Berlin
Phone: + 49 30 2601 2759 · Fax: +49 30 2601 1263 · info@beuth.de · Download: www.beuth.de

3 In-situ measurements of state variables

In-situ measurements can be subdivided into four large groups: the observation of atmospheric state variables such as temperature, moisture, pressure, and wind (this Chapter), the observation of water in liquid or solid form such as precipitation, clouds, and soil moisture (Ch. 4), the observation of abundances of trace gases, aerosols, odours, and radioactivity (Ch. 5), and the observation of fluxes such as radiation, energy, momentum, and mass fluxes, and evaporation (Ch. 6).

3.1 Thermometers

The state parameter internal energy E_I of an ideal gas, which atmospheric air can be considered in good approximation, is a pure function of temperature:

$$E_I = c_v T \tag{3.1}$$

with the absolute temperature T measured in K and the specific heat for constant volume $c_v = 718$ J kg^{-1} K^{-1} at 15 °C. Here, temperature is a measure for the mean kinetic energy of the gas molecules. The Kelvin temperature scale starts at absolute zero and is fixed at two further points. These are the triple point of water at 273.16 K (= 0.01 °C) and the boiling point of water for normal pressure (1013.25 hPa) at 373.15 K (= 100 °C). In thermodynamical relations, the absolute temperature measured in Kelvin must be used.

Temperatures from meteorological measurements are often reported in centigrades (Celsius, °C), denoted with a lowercase t. Especially in North America we still find temperature readings in degrees Fahrenheit (°F). The conversions are as follows:

$$t/°C = T/K - 273.15 \quad \text{or inversely} \quad T/K = t/°C + 273.15 \tag{3.2}$$

$$t/°F = t/°C * 9/5 + 32 \quad \text{or inversely} \quad t/°C = (t/°F - 32) * 5/9 \tag{3.3}$$

Very old-fashioned and no longer in use are temperatures given in degrees Reaumur (°R):

$$t/°R = t/°C * 0.8 \quad \text{or inversely} \quad t/°C = t/°R * 1.25 \tag{3.4}$$

The sensitivity of such a thermometer is inversely proportional to the thickness of the bimetal strip and directly proportional to the square of the length of the strip and to the difference in the thermal expansion coefficients of the utilized metals. In order to realize longer strips, these are often coiled up. The time constant of a bimetal thermometer is about 20 s, i.e. it is faster than a mercury thermometer. The accuracy of a bimetal thermometer is about 0.5 K at its best (Strangeways 2000).

3.1.3 Resistance thermometers, thermistors

The conductivity of electrical conductors usually depends on their temperature. Therefore, the measurement of the conductivity – or its inverse, the resistance – can be employed to measure the temperature. The actual choice of the medium is influenced by, e.g., the long-term stability of the material, the linearity of the resistance change with the temperature change, the size of the resistance change with temperature changes, and the cost of the material. The conductivity of metals decreases with increasing temperature while the conductivity of semiconductors increases with increasing temperature. This dependence on the absolute temperature T can approximately be described by:

$$R(t) = R_0 (1 + \alpha(T - T_0) + \beta(T - T_0)^2 + ...), \qquad (3.8)$$

with the resistance R_0 at temperature T_0. The sensitivity of the thermometer is related to the coefficients α and β characteristic for the utilized metals like copper, nickel, or platinum. For pure platinum, $\alpha = 0.0039$ K^{-1} and $\beta = -5.85 \cdot 10^{-7}$ K^{-2}, for nickel $\alpha = 0.0068$ K^{-1}, and for copper $\alpha = 0.0043$ K^{-1}. Most frequently, a piece of a platinum wire with an electrical resistance of 100 Ω at 0 °C (also called Pt 100) is used. The resistance increases with about 0.4 Ω per degree centigrade. The melting point of platinum is at 1772 °C. Thus Pt 100 thermometers can be employed for a range from –200 °C to +500 °C. Electrical resistance thermometers allow for an immediate electronic post-processing or filing of the measured data. Extremely thin platinum wires with a diameter of 12 μm can be used to measure high-frequency temperature fluctuations necessary for the computation of turbulent heat fluxes (see Ch. 6.3.9).

Instead of metals (so-called PTC or 'positive temperature coefficient' resistors) semiconductors can also be used. Their resistance decreases with increasing temperature, i.e. their α is negative and has an absolute value which is one order of magnitude larger than that of metals. Such sensors are called thermistors or 'negative temperature coefficient' (NTC) resistors. Their advantage is their more rapid adaptation to temperature changes and their larger sensitivity. Unfortunately the dependence on temperature of their resistance is not perfectly linear; therefore platinum sensors are often preferred.

PTC resistors have an accuracy of 0.15 to 0.25 K, NTC resistors an accuracy of 0.05 to 0.5 K. NTC thermometer age and have to be calibrated regularly (Strangeways 2000).

3.1.4 Thermocouples, thermopiles

Thermoelectric elements or thermocouples use the contact voltage between two pieces of different metals touching each other. This effect was discovered by the German physicist T.J. Seeback (1770–1831) in 1821. When two different metals are brought together, electrons transit from one metal to the other until a voltage difference ΔU typical for the participating metals and the temperature of these metals is attained:

$$\Delta U = kT/e \ \ln(n_1/n_2), \tag{3.9}$$

where T is temperature, $k = 1.381 \cdot 10^{-23}$ J K^{-1} the Boltzmann constant, $e = 1.6 \ 10^{-19}$ C the elementary charge of an electron, and n_1 and n_2 the electron densities in the two metals. In thermocouples two pieces of metal are soldered together at one end and are connected to a voltmeter at the other end to form an electric circuit. As long as the soldered contact and the ends with the voltage instrument have the same temperature, no electric current flows. A temperature difference between the two contact points gives rise to a so-called thermoelectric current. If the contact with the voltmeter is kept at a known constant temperature, the other contact point can be utilized as a temperature sensor. For small temperature differences the thermoelectric voltage U_{th} between the two contact points is described by the linear relationship (Gerthsen & Vogel 1993):

$$U_{th} = k/e \ \ln(n_1/n_2) \ \Delta T. \tag{3.10}$$

At larger temperature differences higher order terms enter the equation:

$$U_{th} = a\Delta T + b \ \Delta T^2 \tag{3.11}$$

Table 4. Position of some chemical elements in the thermoelectric voltage sequence at 0 °C. These are relative values with respect to lead whose value was arbitrarily set to 0 (Gerthsen & Vogel 1993).

metal	thermally induced voltage change in µV K^{-1}
antimony	+ 35
iron	+ 16
tin	+ 3
copper	+ 2.8
silver	+ 2.7
lead	0
aluminium	− 0.5
platinum	− 3.1
nickel	− 19
bismuth	− 70

The amount of change of the thermoelectric voltage with temperature determines the sensitivity of the thermocouple. As the ratio k/e of the Boltzmann constant and the elementary charge is just 86 µV K^{-1} the thermoelectric forces are very small. Table 4 lists thermoelectric voltages changes with the voltage for lead set to zero. In order to produce larger voltages, several thermoelectric elements can be concatenated to a thermopile. Thermopiles are widely used e.g. in radiation instruments (see Ch. 6.1).

Because it is quite laborious to keep one of the contact points at a constant temperature, thermoelectric elements are mainly used for the measurement of temperature differences. A display of this difference is principally possible without the aid of an additional power source but most times an electric amplifier is used.

3.1.5 Sonic thermometry

Sound waves with a frequency above the range of human hearing (15 to 20 kHz, for older people often considerably lower) are called ultrasound. Generally, the propagation speed of sound in air, c_s depends only on the temperature of the air:

$$c_s = 20.067 \sqrt{T_a}, \tag{3.12}$$

with the so-called 'acoustic' temperature T_a in K. The acoustic temperature like the virtual temperature (3.5) depends on the humidity of the air (therefore the virtual temperature is often taken as an approximation of the acoustic temperature, see also Kaimal & Gaynor (1991)):

$$T_a = T(1 + 0.513q), \tag{3.13}$$

with the specific moisture q (Raabe et al. 2001). For $t_a = 20\,°C$ ($T_a = 293.15$ K) we get $c_s = 343$ m/s. Therefore ultrasound can be used for measuring the temperature without annoying people.

In contrast to light, sound is transported with moving air. Thus, the travel time of a sound pulse from a emitter to a receiver depends on the temperature and on the component of the wind speed parallel to the sound beam. An ultrasonic anemometer, which can be used to measure the wind as well as the temperature, consists of one to three pairs of ultrasonic transducers with piezoelectric crystals that serve as emitters as well as receivers (see Figs. 28, 55, 56 below for different types and Fig. 17 for a schematic). The transducers operate as emitters as well as receivers. Thus, sound pulses are sent in both directions within each pair of transducers. The two transducers in a pair are mounted about 10 cm apart at thin steel members which exert some minor distortions to the flow. Often, two pairs are arranged horizontally to measure the horizontal wind components and one pair is mounted vertically in order to yield directly the vertical component of the wind. Other constructions operate with three inclined paths in order to minimize the flow deformation by the sensors. The average sound speed computed from signals in both directions is proportional to the acoustic temperature (see (3.12)) the difference between the two sound speeds in the two directions within one pair of transducers is proportional to the wind speed component along this direction. Regarding the temperature, a sonic anemometer yields the temperature fluctuation with high accuracy but the mean temperature can deviate 1 to 2 degrees from the true value, e.g., due to slight deformation of the members bearing the transducers.

The utilized sound frequency is usually around 40 to 200 kHz. 10 to 100 sound pulses are sent per second and the travelling time of the sound pulses in both directions is measured. An ultrasonic anemometer thus permits the recording of high-frequency fluctuations of the temperature and the wind components. The covariance between the temperature fluctuations and the wind fluctuations yields the turbulent heat flux (see Ch. 2.1.5). Therefore, in addition to being a wind and a temperature measuring device, an ultrasonic anemometer is also a heat flux meter.

3.1.6 Measurement of infrared radiation

To be precise, the measurement of the emitted infrared radiation from an object belongs in the class of remote sensing methods (see Ch. 7). The intensity of emitted infrared radiation from a medium depends on the temperature and on the emissivity of the medium. The amount of energy emitted from a black (i.e., with an emissivity of unity) object per unit wavelength interval and per unit area of its surface (the spectral radiative flux density) is given by Planck's law:

$$E(\lambda) = \frac{c_1 \cdot \lambda^{-5}}{exp(c_2/(\lambda T)) - 1}, \qquad (3.14)$$

with the wavelength, λ of the emitted radiation in metres and the temperature, T of the medium in Kelvin. The constants are $c_1 = 1.191 \cdot 10^{-16}$ W m² and $c_2 = 1.44 \cdot 10^{-2}$ m K. This law is named after the German physicist Max Planck (1858–1947) who discovered this relation in 1900.

Integration over all wavelengths, λ yields the total energy radiated from a beamer per unit surface area that depends only on its temperature. The amount of energy is given by the Stefan-Boltzmann law (named after the two Austrian physicists Josef Stefan (1835–893) and Ludwig Boltzmann (1844–1906)):

$$E = \sigma T^4, \qquad (3.15)$$

with the Stefan-Boltzmann constant $\sigma = 5.67 \cdot 10^{-8}$ W/m²/K⁴ (not to be interchanged with the Boltzmann constant in Ch. 3.1.4). Important information is the wavelength at which the maximum energy is irradiated. Differentiating Planck's law with respect to the wavelength yields Wien's law (named after the German physicist and Nobel laureate Wilhelm Wien (1864–1928)):

$$\lambda_{max} T = 0.00289782, \qquad (3.16)$$

with the wavelength, λ_{max} of the maximum irradiance in metres and the temperature, T in Kelvin.

Instruments which determine the temperature of a medium from the amount of emitted thermal radiation, are called pyrometers. Pyrometers usually contain thermocouples or thermopiles to absorb the incoming radiation (see Ch. 3.1.4) and to convert it into temperature information. Also used are bolometers where the received radiation heats the absorbing material and provokes a change in the electrical resistance of this material (see Ch. 3.1.3). The metal bolometer developed by the American civil engineer, astrophysicist and flight pioneer Samuel Pierpoint Langley (1834–1906) consists of blackened platinum stripe or metal coil in an evacuated glass bulb. When this coil is hit by radiation, its electrical resistance changes and can be detected with the help of a Wheatstone bridge. Today, there are also bolometers containing thermistors (Ch. 3.1.3) or even superconductors as a thermometric medium which produce a very large resistance change when this medium converts from a superconductor to a normal conductor by heating.

3.1.7 Soil thermometer

The temperature of the soil can be obtained with mercury thermometers having a right-angled bend between the bulb and the beginning of the scale. The bulb of these thermometers is placed in the designated soil depth, the upper end of the capillary tube with the end of the mercury thread poking out of the soil. This allows for a reading of the instruments without removing them. Such thermometers can be used for soil depths up to 1 m. For greater depths, thermometers can be fixed permanently in drilled holes from which they can be lifted for inspection. More and more resistance thermometers are used for the measurement of soil temperatures because such thermometers can be placed in any depth and connected with the data logger by electrical wires (Strangeways 2000).

3.1.8 Recommendations for temperature measurements

Liquid-in-glass thermometers are used for precise measurements in screens where they are read by observers. Principally, alcohol thermometers are suited for somewhat lower and mercury thermometers for somewhat higher temperatures. Nowadays, routine temperature observations at automated weather stations are performed with resistance thermometers. Minimum and maximum temperatures are determined from the temperature time series. High-frequency measurements, especially for flux measurements, are made with ultrasonic anemometers or very thin resistance thermometers. Precise measurements of small temperature differences, e.g., in radiation instruments, are often executed by employing thermocouples or thermopiles. Infrared sounding is used for example, for soil temperature measurements from a mast or for temperature measurements of inaccessible objects.

3.1 Thermometers

Fig. 16. Radiation shield for temperature measurements (temperature hut).

With the exception of ultrasonic and infrared temperature measurements, it is always important to protect the sensor from incoming radiation that could also heat up the temperature sensor (Fig. 16). Radiation-shielded thermometers usually have to be ventilated. A permanent air flow ensures that the sensor adapts to the air temperature and not vice versa. For thermometers outside a screen (Fig. 8), the radiation shield usually consists of 12 cm long tubes with a diameter of about 2 cm that are

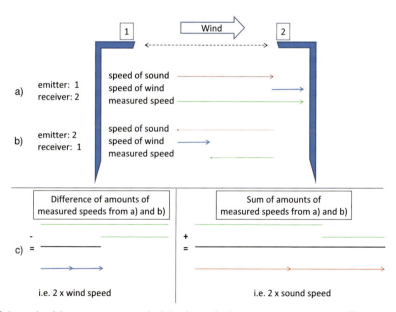

Fig. 17. Schematic of the measurement principle of a sonic thermometer-anemometer. The measurement is made along the dashed line between the numbers "1" and "2" in square boxes. The upper parts a) and b) of the Figure show the measured speeds, the lower part (c) shows the derivation of wind speed and sound speed (which is proportional to the temperature) from the measured speeds in a) and b). The lengths of the arrows in a) and b) and of the lines in c) are proportional to the speeds. (See also Fig. 28)

rology, surface pressure is reported in hPa because this unit is identical the formerly used millibar (mb). 1013.25 hPa is the surface pressure in a standard atmosphere and refers to a mercury column of 760 mm height or a weight force of 10335.15 kg/m². This standard value was in former times also given as 760 Torr (in memory of Evangelista Torricelli (1608–1647) who invented the mercury barometer).

The atmospheric pressure decreases with height. Therefore, if the surface pressure value is known, barometers can also be used as altimeters.

Table 6. Overview on pressure sensors.

device	sensor	principle	transformer / amplifier	display
barometer	mercury column	balances the weight of an air column	evacuated tube	analogue: height of the mercury column in the tube
aneroid barometer	evacuated capsule with a spring inside	equilibrium between air pressure and spring force	mechanical via a lever	analogue: pointer
hypso-meter	water	boiling point depends on air pressure	see the respective thermometers	see the respective thermometers
electronic pressure transducers	silicon diaphragm	air pressure deforms the diaphragm	electric capacity or resistance of the transducer is changing	digital in file or to a digital display
microbaro-meter	dewar flask	comparison of inside and outside pressure	balancing between inside and outside pressure by a small valve	
pressure balance	piston	comparison of weights on a balance		analogue or digital

3.3.1 Barometers

Six different methods for the determination of the air pressure are in use (Tab. 6): the mercury barometer, the aneroid barometer, the hypsometer, electronic pressure sensors, microbarometers, and pressure balances.

3.3.1.1 Mercury barometers

Barometers compare the weight of a column of air with the weight of a column of liquid of the same horizontal cross section. The liquid is usually mercury, because it has a very high specific weight of 13595.1 kg/m³. This allows for a compact and still movable instrument. A water barometer would have a column height that is about 13.5 times that of a mercury barometer, and thus would be really impractical. Actu-

3.3 Pressure sensors

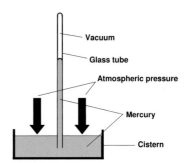

Fig. 20. Schematic of a mercury barometer. The air pressure (black arrows) drives the mercury into the evacuated glass tube.

ally, a water barometer was constructed by Otto von Guericke (1602–1686), and in 1660 he indeed used it for the forecast of an approaching storm. In addition to its high specific weight, mercury has a very low vapour pressure, i.e. losses by evaporation are negligible.

Mercury barometers consist of an evacuated glass tube that is turned upside down. The lower open end of the tube resides in a pool or cistern of mercury (Fig. 20). The atmospheric pressure drives mercury into the evacuated tube until the weight of the mercury column inside the tube equals the weight of the air column above the site having the same cross-section as the tube. However, the measurement principle is sensitive to the temperature of the instrument because mercury and glass have different thermal expansion coefficients. Therefore, a correction of the barometric readings with respect to temperature must be applied (see also Ch. 2.3.2). For this purpose usually a small thermometer is attached to the outside of the barometer housing. Furthermore, pressure measurements depend on the exact value of the Earth's gravity force at the site. The latter is varying with latitude and altitude above sea level. Surface pressure values are usually corrected to a standard gravity of 9.80665 m/s². The accuracy of a mercury barometer is about 0.1 hPa.

3.3.1.2 Aneroid barometers

Aneroid (greek: without liquid) barometers consist of a small evacuated metal capsule or box with a spring inside that prevents the capsule from collapsing (Fig. 21).

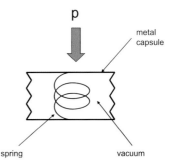

Fig. 21. Schematic of a Vidie capsule (aneroid barometer). The spring inside the capsule works against the air pressure (grey arrow).

Fig. 22. Mechanical barograph constructed by Jules Richard in Paris in 1898, based on a Vidie capsule (upper right). This large evacuated capsule does not contain a spring as in Fig. 21. Rather, the compressing force of the air pressure is counteracted by a weight of about 126 kg fixed to the pole underneath the capsule.

This instrument was invented by Lucien Vidie (1805–1866) in 1843 and is sometimes named after him. Data from aneroid barometers do not require a temperature and gravity correction. But the instruments have to be calibrated against a standard barometer. Key advantage of aneroid barometers is their small size, which make them well suited for mobile measurements. The deformation of the metal capsule can be recorded mechanically or electrically (via a resistance change or by induction). Mechanical translation of the compression of the metal capsule has led to the construction of barographs (Fig. 22) similar to thermo- and hygrographs. Their accuracy is some tenths of a hPa (Strangeways 2000).

3.3.2 Hypsometers

Hypsometers rely on the observation that the boiling point of a liquid depends on ambient air pressure. A liquid boils when its temperature-dependent vapour pressure equals the ambient air pressure. For near-surface measurements, this temperature dependence can approximately be described by:

$$t_S = 100\ °C + 0.02766\ °C\ (p - 1013.2\ hPa) - 0.0000124\ °C\ (p - 1013.2\ hPa)^2, \qquad (3.29)$$

where p is the ambient air pressure in hPa. At an ambient pressure of 701 hPa, water boils at 90 °C, at 473.4 hPa at 80 °C.

In a hypsometer (greek: altimeter, the name indicates that the instrument was originally developed for altitude measurements in the mountains), the boiling point of distilled water is determined very precisely. A change of 0.028 °C corresponds to a pressure change of 1 hPa. Therefore, the temperature measurement must be accurate to at least 0.01 °C. The liquid in a hypsometer is heated either electrically or by

3.3 Pressure sensors

a small alcohol burner. In former times, hypsometers were used as standard instruments where mercury barometers could not be deployed. The measurement principle of a hypsometer is an absolute one and requires no calibration. Today, the pressure balance (Ch. 3.3.5) is more commonly used as standard instrument.

3.3.3 Electronic barometers

The barometers introduced above are not suited for integration into automated measurement setups. For this purpose, electronic barometers that make use of transducers which produce a pressure-related electric quantity (see e.g. VDI guideline 3786, part 16, see Appendix) have been developed.

3.3.3.1 Capacitive pressure transducers

Capacitive pressure transducers consist of a flexible very thin silicon diaphragm with areal size A as a sensor over a cavity of a carrier substance separated by an insulator. Electrodes are attached to the diaphragm and the carrier substance (Fig. 23). Pressure-induced deformations of the diaphragm results in a change of the electrical capacity of the sensor element, which is then recorded and converted electronically into atmospheric pressure. A reduction of the distance between the silicon diaphragm and the carrying material Δd leads to an increase of the electric capacity ΔC:

$$\Delta C = \varepsilon_0 \varepsilon_\tau \frac{A}{\Delta d}, \tag{3.30}$$

with the electric field constant, $\varepsilon_0 = 8{,}854 \cdot 10^{-12}$ AsV^{-1}m^{-1} and the dielectric constant, ε_τ (for air = 1). The accuracy of these sensors is about a few tenths of a hPa (Strangeways 2000).

3.3.3.2 Resistive pressure transducers

Similarily, strain gauges forming a Wheatstone bridge can be bonded to a silicon diaphragm of a pressure transducer. The pressure-induced deformation of the diaphragm

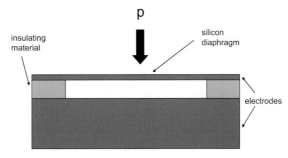

Fig. 23. Schematic of a capacitive pressure sensor. The distance between the flexible silicon diaphragm and the solid lower electrode is altered by the air pressure (black arrow). This changes the electrical capacity of this sensor.

now produces a length change, Δl and simultaneously a change in cross-section, ΔA of the strain gauges, which results in a change of the electrical resistance, ΔR:

$$\Delta R = \rho \frac{\Delta l}{\Delta A}, \tag{3.31}$$

Here, ρ is the specific resistance of the substance of the strain gauges. A stretching with a simultaneous reduction of the cross-section leads to an increase of the resistance. These pressure sensor elements are called piezo-resistive pressure transducers. Their accuracy is about 0.3 hPa (Strangeways 2000).

3.3.3.3 Quartz pressure transducers

The oscillation frequency of a quartz crystal resonator decreases if a force is exerted on the crystal. By means of a system of levers, the atmospheric pressure is translated to the crystal, which itself is kept in a vacuum. These transducers also provide temperature information because their resonance frequency is a function of temperature, too.

3.3.4 Microbarometer

Microbarometers are used for monitoring short-term high-frequency pressure fluctuations where the absolute value of the pressure is of secondary importance. For this purpose, piezo-electric pressure transducers and Dewer flasks are well suited. The latter measure small high-frequency ambient air pressure fluctuations in comparison to a slowly changing pressure in the flask. Through a small valve, the interior of the flask stays in contact with the ambient air and the pressure inside the flask slowly adapts to the ambient air pressure. By this adaptation, low-frequency pressure fluctuations are eliminated from the pressure records. The Dewar flask is well insulated in order to minimize disturbances of the measurements from outside temperature fluctuations. A practical example of a Dewar flask microbarometer is demonstrated in Emeis & Kerschgens (1985).

3.3.5 Pressure balance

In a pressure balance (a type of a piston manometer, also called deadweight tester or deadweight gauge), the pressure exerts a force on a piston that is compared to the weight of gauged weights. For atmospheric pressure measurements, the weights for comparison have to be in a vacuum in order to avoid contamination by dust. The atmospheric pressure exerts a force to the lower surface of the piston which resides in a vertical cylinder and which is balanced by the counterweights. The piston is kept rotating to minimize the friction between the piston and the wall of the cylinder. Piston manometers register the absolute pressure and are used as standard instruments today (Woo et al. 2004, WMO 2006, VDI guideline 3786, part 16 (see Appendix)).

3.3.6 Recommendations for pressure measurements

Piston manometers and other manometers feature the lowest measurement errors and are therefore used as standard instruments (Woo et al. 2004). Automated measurements are best conducted with aneroid barometers and electronic pressure transducers, because they are best suited for an electronic post-processing of the data.

Pressure measurements are usually representative for a larger area if they were taken at a distance from flow obstacles where larger dynamic contributions (a few per mils) to the total pressure can occur. Atmospheric pressure measurements can also be made inside buildings. The altitude of the measurement site must known with high accuracy because an 8 m difference in altitude introduces an error of 1 hPa. In order to determine a horizontal pressure distribution in orographically structured terrain, the pressure values have to be referenced to a standard surface (usually the sea level). For this purpose, the air temperature must be determined.

Technical rules for pressure measurements are documented in the VDI guideline 3786, part 16 (see Appendix) and in Ch. 3 of WMO (2006).

3.4 Wind measurements

In contrast to other meteorological variables, winds are vectors, i.e. characterized by an amount (wind speed) and a direction (wind direction). Generally, (apart from small-scale and convective processes) the vertical wind component is much smaller than the horizontal wind components. Therefore, frequently, only horizontal wind speed is measured. Furthermore, wind is a highly variable atmospheric parameter. Its speed and direction fluctuate strongly. This phenomenon is called the gustiness of the wind.

In the following, x and y are the two horizontal directions (positive towards the East and the North), z is the vertical direction (positive upward), and $u(t)$, $v(t)$, and $w(t)$ are the three time-dependent wind components in the three directions x, y, and z, allowing following wind speeds to be defined:

the mean wind speed U:

$$U = T^{-1} \int_0^T \sqrt{(u^2(t) + v^2(t) + w^2(t))}\, dt, \qquad (3.32)$$

the mean horizontal wind speed:

$$U_h = T^{-1} \int_0^T \sqrt{(u^2(t) + v^2(t))}\, dt, \qquad (3.33)$$

and the mean vertical velocity:

$$W = T^{-1} \int_0^T w(t)\, dt, \tag{3.34}$$

Additionally, the following parameters can be obtained:

the mean wind direction:

$$\Phi = T^{-1} \int_0^T \arctan(u/v)\, dt \tag{3.35}$$

and the variances of the wind components:

$$\sigma_u = \sqrt{T^{-1}(\int_0^T (U - u(t))^2\, dt)} \tag{3.36}$$

$$\sigma_v = \sqrt{T^{-1}(\int_0^T (V - v(t))^2\, dt)} \tag{3.37}$$

$$\sigma_w = \sqrt{T^{-1}(\int_0^T (W - w(t))^2\, dt)} \tag{3.38}$$

From (3.36) to (3.38) the turbulence intensity, σ_u/U (frequently given in %) and the turbulent kinetic energy, $q = 0.5\, \rho\, (\sigma_u^2 + \sigma_v^2 + \sigma_w^2)$ can be computed.

Due to the gustiness of the wind, there is nearly always the necessity to determine mean values for the wind. This involves some difficulties, because the different available averaging procedures can lead to different results. Therefore, the averaging procedures must be selected according to the purpose of the measurement. We must distinguish between vector mean, component means, and scalar mean. The vector mean can be visualized by concatenating several vectors to a longer chain by preserving their orientation. The mean is then a vector pointing from the end of the first vector to the tip of the last vector and whose length is divided by the number of vectors involved in the mean. For example, the vector mean of two vectors with the same length but opposite directions is the zero vector. Mathematically the vector mean can be computed by separating the vectors in their components (e.g. the horizontal wind into the East and the North component), computing separate means for the two or three components, and finally forming the resulting vector from the means of the two or three components. A scalar mean is computed by averaging the lengths of the vectors disregarding their orientations. A scalar mean wind speed is always larger or equal (if all wind vectors had the same orientation) than the vector mean wind speed.

Some measurement systems which only determine scalar mean wind speeds (e.g. cup anemometers) separately observe the wind direction with additional sensors (e.g. wind vanes). In the case of such an observation, the average over two opposing wind measurements with equal wind speeds leads to a mean wind speed of just this speed and an undefinable wind direction. It can be dangerous, and often even meaningless, to compute scalar averages from wind direction data. This is permissible only for data coming from a narrow wind direction sector. (Imagine, you have two wind readings

3.4 Wind measurements

with equal wind speed but opposite directions, say 90° and 270°. The average (vector) wind is vanishing, but separate scalar averaging of wind speed and wind direction would give a wind coming from 180°. This latter result is certainly wrong.)

Different methods for wind measurements are listed in Table 7. Wind data are reported in a variety of units (Tab. 8); full compliance to international standards has not

Table 7. Overview on wind measurement instruments.

device	sensor	principle	transformer / amplifier	display
wind vane	solid blade	moves in a position with lowest flow resistance	electrical contacts	analogue or digital
wind sock	textile tube	moves in a position with lowest flow resistance	–	–
pressure plate	solid blade	comparison of wind force to gravity	–	analogue: observation of a deviation from rest from a given scale
cup anemometer	three to four rotating cups mounted at a vertical axis	cups are driven by the moving air		analogue and digital
aerovane	wind vane with a propeller fixed to the other end	wind vane turns propeller into the wind direction, propeller rotates proportional to wind speed	electrical contacts	digital in file or on a display
pressure tube	tube with a small opening at its tip facing the wind	measurement of the pressure head at the tip of the tube		digital in file
hot wire anemometer	thin wire	wind speed dependent heat loss	depends on the used thermometers and current meters	digital in file
sonic anemometer	sound transducer	sound waves are moved by the wind	electronic amplification	digital in file
propeller anemometer	three propellers	rotation speed of the propeller	electrical contacts	digital in file

Table 8. Conversion table for wind sped units.

	km/h	ft/s	mph	knots	m/s
km/h	1	0.912	0.621	0.540	0.278
ft/s	1.096	1	0.680	0.592	0.305
Mph	1.611	1.470	1	0.870	0.447
knots	1.852	1.689	1.150	1	0.514
m/s	3.600	3.281	2.237	1.944	1

Table 9. Beaufort scale for wind speed (stages 13 to 17 added later) (Strangeways 2003, Wikipedia).

Wind force	name	onshore effects	offshore effects	wind speed in 10 m height
0	calm	smoke rises vertically	flat sea	0.0–0.2 m/s
1	light air	smoke motion visible	ripples without crests	0.3–1.5 m/s
2	light breeze	wind felt on face	small wavelets	1.6–3.3 m/s
3	gentle breeze	leaves and smaller twigs in motion	large wavelets, crests begin to brake, some whitecaps	3.4–5.4 m/s
4	moderate breeze	small branches move, dust and loose paper raised	small waves, whitecaps	5.5–7.9 m/s
5	fresh breeze	moderate branches move, smaller trees begin to sway	moderate waves, whitecaps all over	8.0–10.7 m/s
6	strong breeze	large branches move, umbrellas hard to move	large waves with foam crests	10.8–13.8 m/s
7	near gale	whole trees move	foam begins to be blown in streaks	13.9–17.1 m/s
8	gale	twigs broken from trees, cars veer on roads	moderately high waves, breaking crests, streaks of foam	17.2–20.7 m/s
9	strong gale	stronger branches break off, slight structural damages	high waves, crests start to roll over, spray	20.8–24.4 m/s
10	storm	trees uprooted, first damages to roofs	very high waves, spray reduces visibility	24.5–28.4 m/s
11	violent storm	widespread vegetation damage, structural damages	exceptionally high waves, foam covers much of the sea surface, reduced visibility	28.5–32.6 m/s
12	hurricane	considerable and widespread damage to vegetation and structures	huge waves, sea is completely white, greatly reduced visibility	32.7–36.9 m/s
13	hurricane	considerable and widespread damage to vegetation and structures	huge waves, sea is completely white, greatly reduced visibility	37.0–41.4 m/s
14	hurricane	considerable and widespread damage to vegetation and structures	huge waves, sea is completely white, greatly reduced visibility	41.5–46.1 m/s
15	hurricane	considerable and widespread damage to vegetation and structures	huge waves, sea is completely white, greatly reduced visibility	46.2–50.9 m/s
16	hurricane	considerable and widespread damage to vegetation and structures	huge waves, sea is completely white, greatly reduced visibility	51.0–56.0 m/s
17	hurricane	considerable and widespread damage to vegetation and structures	huge waves, sea is completely white, greatly reduced visibility	> 56.0 m/s

everywhere been achieved yet. The SI unit for wind speed is m/s. In daily life often also km/h is used, especially in Europe. Predominantly in the USA and in some other countries, miles per hour (mph) or feet per second (ft/s) are still used today. In many marine and military institutions, knots (nautical miles (1852 m) per hour) are still standard. Estimates of the wind speed in Beaufort (Tab. 9) are still common practice.

3.4.1 Estimation from visual observations

Observing rising smoke columns, the movement of trees, the formation of sea waves, or wind damages can be used to estimate wind direction and speed (see the definitions of the Beaufort scale in Tab. 9).

Wind socks are often used as optical indicators of wind speed and direction (Fig. 24). A wind sock is a textile tube with red and white rings attached to a mast at a height of a few metres. Both ends of the tube are open and the tube is moved by the wind flow streaming through it. Wind socks are often found at airports, helicopter landing sites, motorway bridges and other places prone to strong winds.

Another example of a non-quantifying wind instrument is the pressure plate designed by the Swiss physicist and meteorologist Heinrich von Wild (1833–1902). This is a metal table whose upper edge is hung to a bearing boom that is moved in a position perpendicular to the wind direction by a wind vane attached to the boom. The wind pushes the table from its resting position (vertically downward). The degree of deviation (angle) is a measure of the wind speed. A scale next to the table allows reading the wind speed.

3.4.2 Wind direction

The wind direction can be determined using a wind vane. This is a horizontally rotating solid vertical blade (mounted to a vertical axis of rotation). The wind moves the blade into the position where it exerts the minimal flow resistance. The blade is pointing to the opposite direction from where the wind is coming. In meteorology, the wind direction is the direction from where the wind blows. This definition is opposite to the definition of the direction of ocean currents. The position of the turnable blade can be registered by reading electrical contacts mounted circularly around the rotating vertical axis.

Fig. 24. Wind sock indicating wind direction and strength.

3.4.3 Cup anemometer

The cup anemometer is a classical and robust wind speed sensor (Fig. 25). It usually has three, sometimes four, cups that are mounted like a star to a vertical turnable axis. Because the flow resistance is much larger for a flow approaching the open side of the cups than the closed side of them, the wind is driving the cups always in the same direction regardless from where the wind is coming. The revolution speed of the cups is thus proportional to the wind speed. Considering the distance of the cup centre from the rotating axis, the path of the wind per unit time can be computed.

Disadvantages of a cup anemometer are the non-zero minimum wind speed necessary to start the instrument turning (about 0.3 to 0.5 m/s), and the tendency of the instrument to record too high wind speeds if the wind is very gusty. This latter effect is called overspeeding (Busch & Kristensen 1976) or u-error (MacCready 1966). It occurs because the instrument speeds up faster than it slows down. The overspeeding depends strongly on the construction of the instrument. For some instruments an overspeeding of more than 7 % at a turbulence intensity of 23 % can be observed (Pedersen 2003) and should be determined by calibration in a wind tunnel. MacCready (1966) calls this the 'u-error' and gives a rough estimate of the overspeeding as $\Delta u = 1/20 \, d/z$, where z is the measurement height and d the distance constant. d is defined as the distance the wind must travel until the instrument has followed up a 63.2 % of a step change in wind speed. Busch & Kristensen (1976) derive a more complex relationship which also takes into account the surface roughness length and atmospheric stability via the Monin-Obukhov length. An extensive discussion on the biases or errors of a cup anemometer can be found in Kristensen (1993).

The inertia of cup anemometer (and thus overspeeding) is minimized by reducing the distance between the centre of the cups and the rotating axis. The inertia is usually characterized by the distance constant d. If the distance constant is known the temporal resolution, τ of a cup anemometer can be determined by:

$$\tau = d / u. \tag{3.39}$$

For $d = 2$ m (a quite good distance constant) and a wind speed $u = 5$ m/s, we get $\tau = 0.4$ s.

Fig. 25. Cup anemometer measuring wind speed (left) and wind vane indicating wind direction (right) (Photo: Helmut Mayer).

3.4 Wind measurements

Cup anemometers are calibrated for horizontally upcoming flow. If the flow approaches at an angle to the horizontal the measured wind speed can – depending on the construction of the instrument – deviate from the true wind speed (w-error in the terminology of MacCready 1966). The w-error is due to turbulent vertical components of the wind which influence the measurement of the horizontal wind speed and it increases with unstable atmospheric stratification. According to MacCready, an overestimation of the horizontal wind speed by 10 % is probably not uncommon. For instruments designed for the measurement of horizontal wind speeds only, the measured wind speed ideally is reduced proportionally to the cosine of the angle between the wind direction and the horizontal. For instruments designed for the observation of the three-dimensional wind speed, ideally no reduction of the measured wind speed should take place. The exact behaviour of cup anemometers has to be determined in wind tunnels. For a given turbulence intensity a cup anemometer can be calibrated to an accuracy of better than 1 %.

MacCready (1966) describes another error, the DP-error. The DP-error appears only when a time average is computed. If there are wind direction fluctuations, then the vector average will give a lower wind speed than a scalar average (see Ch. 8.3.3 below). This error can reach 10 % of the mean speed if the variance of the wind direction is greater than 30° (which is not uncommon according to MacCready). The word 'error' for these effects may be misleading, especially for the DP-error. It depends very much on the application for which the mean wind speed has to be determined as to whether a scalar or a vector average should be formed. If the wrong average is formed, then an error is produced.

3.4.4 Pressure tube

A flow coming onto an obstacle must flow around it. Upstream of the obstacle, a point is found where the streamlines divide and the wind speed is zero (the stagnation point). The pressure p at this point is formed from the static pressure p_s (see (3.26)) which would prevail in a flow at rest, and from the dynamic pressure p_d (see (3.27)). If the dynamic pressure is known, the flow speed can be determined by:

$$v = \sqrt{(2\,p_d/\rho)}. \tag{3.40}$$

The pressure tube (Fig. 26) measures – if constructed as a Pitot tube – the total pressure p, or – if it is constructed as a Prandtl tube – the dynamic pressure p_d. The pressure tube dates back to an invention in 1732 when the French physician and technician Henry Pitot (1695–1771) developed it to measure the flow speed in flowing rivers (Strangeways 2003).

In order to understand the principle of a flow measurement with a pressure tube (see also Fig. 27), imagine first a tube which is closed at the upwind end but which has openings on its side. The interior of the tube is connected to a manometer. Such an instrument measures the static pressure p_s. If the side openings are closed and the therefore are well suited to measure atmospheric turbulence. Using covariance with simultaneously determined temperature fluctuations (see Ch. 3.1.5) permits computing the turbulent heat flux (see Ch. 6.3.9). Thus the ultrasonic anemometer is simultaneously a wind, temperature, and heat flux instrument.

The lack of movable parts in ultrasonic anemometers is very advantageous as this reduces maintenance efforts. It permits the gathering of data from all three wind speed components with high temporal resolution. The scalar mean as well as the vector mean of the wind speed can be determined. Although the principle works very well, it is less reliable in rainy conditions when water on the sensor changes the acoustic path length and therefore the calibration. This makes sonic anemometers less suitable as all-weather instruments (WMO 2006).

3.4.7 Propeller anemometer

Fig. 29. Three-component propeller anemometer (Photo: Helmut Mayer).

Propeller anemometers are simple instruments for the simultaneous determination of all three wind components. It consists of three small propellers with a diameter of a few centimetres mounted perpendicular to each other (Fig. 29). The rotation speed of the propellers is proportional to the wind speed component perpendicular to the propeller plane. This instrument can be used for some turbulence studies, it temporal resolution is about 5 to 10 Hz.

There are instruments combining a propeller for wind speed and a vane for wind direction measurements called aerovanes (Fig. 30). These instruments are, however, unsuitable for turbulence measurements. Propeller anemometers may be subject to the so-called v-error (MacCready 1966). The v-error

Fig. 30. Aerovane, a combination of a propeller anemometer measuring wind speed (right) with a wind vane indicating wind direction (left).

results from azimuthal variations in the wind (wind direction variance) which could lead to misalignments of the measuring device, i.e. the plane of the propeller is not always perpendicular to the wind direction.

3.4.8 Recommendations for wind measurements

The classic instrument for measuring horizontal wind speed is the cup anemometer that can be calibrated to an accuracy of 1 %. But this accuracy depends on the turbulence intensity. The measured speeds increase with increasing turbulence intensity (overspeeding) even if the true mean wind remains constant. The respective wind direction is most frequently determined with a wind vane. In cases where all three components of the wind are required to be known separately, ultrasonic or three-dimensional propeller anemometers must be used. The wind direction can then be computed from the measured wind components. Turbulent wind fluctuations can be observed using hot wire anemometers with thin wires, pressure tubes (five-hole-sondes), or ultrasonic anemometers. Aboard aircraft only pressure tubes and five-hole probes are used. The values from all instruments can be filed and post-processed electronically. For ultrasonic anemometers see also Chapter 3.1.8.

Flow obstacles such as buildings or trees can exert considerable influence on wind measurement data. Therefore, the choice of a site that is representative for its surroundings needs considerable attention. Distances of several times the height of an obstacle are necessary to rule out unwanted side effects on the measurements. On roof-top sites, speed-up effects due to the deviated flow over the building must be considered. Wind measurements in orographically structured terrain with large changes in elevation and surface characteristics (valleys, slopes, hill tops, coast lines, etc.) generally have a low spatial representativity.

Further, for evaluation of wind data after the measurement campaign, the exact altitude of measurement above ground must be known, because wind speed increases strongly with height in the surface layer. Over horizontally homogeneous terrain the measurements may be corrected for the instrument height if the surface roughness of the underlying surface and the thermal stratification of the air in the surface layer are known (by using a logarithmic law or power law for the vertical wind speed increase). Such corrections should be made only for small altitude intervals of up to a few tens of metres. To assess wind energy potential at the hub height of modern wind converters (around 100 m above ground), the use of remote sensing methods is recommended (see Ch. 8.3.1).

Technical rules for wind measurements are documented in the VDI guideline 3786, part 2 (general wind measurements) and part 12 (sonic anemometers), and in Chapter 5 of WMO (2006).

4 In-situ methods for observing liquid water and ice

This Chapter describes in-situ methods for rain, snow, and ice and for soil moisture. The measurement of gaseous water in the atmosphere (atmospheric humidity) is covered in Ch. 3.2. An extended description of instruments for measuring precipitation can be found in the monograph by Strangeways (2007). Modelling of hydrological processes in small catchment areas is described in Plate & Zehe (2008).

4.1 Precipitation

Apart from the aggregate state of the precipitation elements (rain, snow, hail, etc.), and the type of the precipitation events (continuous rain, shower, etc.), the drop size, the amount, the intensity, and the duration of the precipitation must be documented. The amount is usually reported in liters per square metre (l/m^2) or in mm, both denoting the same amounts. General precipitation measurement (Tab. 10) devices are termed hyetometers. Rain gauges are called ombrometers (from the greek word for rain) or pluviometers (from the latin word for rain).

Table 10. Overview on precipitation measurement instruments.

device	sensor	principle	transformer/ amplifier	display
rain sensor	sensitive surface	electrical conductivity	electronics	digital
rain gauge (totalisator)	funnel with container underneath	determination of an amount of liquid	shape of the collecting container	analogue
recording rain gauge	siphon, tipping bucket	determination of an amount of liquid and the number of operations per unit time	volume of siphons, volume of the tipping buckets	analogue/ digital
disdrometer	styrofoam body	transmission of the momentum of a drop provokes a displacement of the body	electronics	digital
disdrometer	light beam or band	determination of the drop size spectrum and the drop fall speed by light absorption	electronics	digital

4.1 Precipitation

4.1.1 Rain sensors (Present Weather Sensors)

Rain sensors provide a signal as to whether it rains or not. These sensors rely on the electric conductivity of rain drops. They consist of a non-corrosive (e.g. gold coated) surface covered with conductor paths at narrow distances of 0.5 mm to each other (Fig. 31). Rain drops on this surface lead to an electric current between adjacent conductor paths and thus to a signal. The surface of the sensor is inclined about 15 degrees to make the droplets run off fast. When a rain drop has been counted, the surface is heated to dry it rapidly in order to prepare it for the detection of the next droplet. Fog and dew do not provoke a signal. Rows of little sticks on the surface help to keep solid precipitation like snow and hail a little bit longer and to improve the detection of this type of precipitation. A short description of such a sensor is given in Plaisance et al. (1997).

Fig. 31. Electrical rain sensor. Rain drops lead to electrical currents between adjacent conductor paths on the surface of the instrument. The rows of little sticks on the surface help to keep solid precipitation like snow and hail a little bit longer and to improve the detection of this type of precipitation.

Fig. 32. Present weather sensor. Pressure, temperature and humidity sensors are mounted underneath the white radiation shield (analogous to the one depicted in Fig. 16), wind speed and direction is measured acoustically with ultrasound from the three sensors at the top of the instrument (see also Figs. 17 and 28 for comparison) and liquid precipitation and hail is detected by the silvery impact plate.

Another type of rain sensor produces a flat beam of light of a few centimetres length and an area of 25 cm². A predefined number (1 to 15) of interruptions in a 50 s interval yields the signal 'begin of precipitation'. Likewise missing interruptions for a selectable period of 25 to 375 s yields the signal 'end of precipitation'. The instrument registers droplets with a diameter of 0.2 mm or larger.

There are further optical methods (e.g. the analysis of the intensity of forward scattering) and instrument which also deliver temperature information. Latter devices are also called 'Present Weather Sensors', if they are able to send automated messages using the WMO code table 4680. Modern Present Weather Sensors are able to measure wind speed and direction, pressure, temperature and humidity, liquid precipitation and hail. The latter is measured acoustically by sensoring the impact of the drops on a plate on top of the instrument (Fig. 32).

4.1.2 Rain gauges (totalisators)

Collecting or integrating rain gauges (totalisators, Fig. 33) consist of a collecting funnel of known area on the top of the instrument having a hole in the middle and underneath a container in the inside of the instrument (Fig. 36). At climate stations these instruments are read daily and the water from the container is poured off to reset the instrument. The container collecting the water has an engraved scale which gives the daily precipitation sum.

Instruments used in Germany typically have a collecting funnel with a straight surface of 200 cm² (a diameter of 15.96 cm). In regions which are difficult to access, the precipitation can also be collected for longer time periods before the instruments are read and reset. The German Weather Service utilizes instruments which were originally constructed by the German meteorologist Gustav Hellmann (1854–1939). To detect snow and hail, these rain gauges can be heated, but the heating produces a non-quantifiable evaporation error. To measure snow under windy conditions, a so-called snow cross is inserted into the collecting funnel to prevent the snow from being blown out of the funnel again.

Fig. 33. Rain gauges or totalisators for accumulative measurement of precipitation.

4.1.3 Pluviographs

Pluviographs or chart-recording rain gauges may be used to determine the duration and intensity of precipitation events. Different designs are available. One type uses an intermediate container with a siphon tube. On the water in this intermediate container a float rises with the water level that is connected via a lever to an arm holding a pen which writes on a stripe of paper mounted on a rotating drum. When a given level of water is reached in the intermediate container, the water is emptied through the siphon tube into a collecting container underneath.

Fig. 34. Wind shields for rain gauges.

In other types of pluviometers, the water drips from the collecting surface into a small tipping bucket (Fig. 35). When the bucket to the one side is filled, the tipping bucket turns over and the other bucket is filled. Each turn over refers to a known amount of precipitation. The number of turnings per unit time is counted.

Alternatively, the mass of the collected rain in the container may be weighed and registered continuously. When the

Fig. 35. Tipping bucket to the inside of a rain gauge (Photo: Helmut Mayer).

container is filled, it must be emptied either manually or via a siphon. For small amounts of rare rain events the evaporation from the collecting container may be sufficient. Time-resolved precipitation monitoring can also be performed with a rain RADAR that is described in Chapter 7.2.1.1.

4.1.4 Disdrometer

Disdrometers – also called drop spectrographs – can capture the size spectrum of rain drops, from which the formation processes of rain can be inferred and which is also necessary for the calibration of rain RADARs. The original disdrometer developed by Joss & Waldvogel (1967) has a styrofoam body, that is kept in its position by two magnetic coils, onto which the rain drops impact. The impacting drops transfer their momentum to the styrofoam body, thereby deviating it slightly from its reference position. This deviation induces an electric current by magnetic induction in another pair of coils. From the intensity of this current, the size of the drops can be estimated. A disdrometer is especially suited to monitor large rain drops. The

Table 11. Conversion factors from volumetric to gravimetric units for trace gases and vice versa for two different ambient temperatures (Colls 2002).

gas	molecular weight in g	ppb to µg/m³ 0 °C	ppb to µg/m³ 20 °C	µg/m³ to ppb 0 °C	µg/m³ to ppb 20 °C
SO_2	64	2.86	2.66	0.35	0.38
NO_2	46	2.05	1.91	0.49	0.52
NO	30	1.34	1.25	0.75	0.80
O_3	48	2.14	2.00	0.47	0.50
NH_3	17	0.76	0.71	1.32	1.43
CO	28	1.25	1.16	0.80	0.86

gravimetric units relies on the fact that one mol of an ideal gas, which has 6.02 10²³ molecules, weighs M g and takes a volume of 0.0224 m³. Here, M is the molecular weight. Table 11 gives some conversion factors.

Many methods need an accumulation step ahead of the real measurement step itself, because otherwise the very low concentrations cannot be quantified reliably. For gaseous concentrations, e.g., this can be achieved by the operation of a denuder, while for aerosols this can be done by collecting them on a filter paper. This Chapter describes in-situ methods for accumulation and measurement methodologies. The first half is devoted to trace gases, the second to aerosols. A detailed description of many methods can be found in Colls (2002). Table 12 gives an overview of the methods as sorted by trace species.

5.1 Measurement of trace gases

Basically, measurement methods can be divided into physical and chemical methods. Some of the methods offer a continuous monitoring of the gas concentrations; others require sampling (Fig. 39 shows stainless steel containers for gas samples) and a later analysis in the laboratory. Table 13 gives an overview on the available techniques to quantitatively detect trace gases.

Fig. 39. Stainless steel containers for the transport of gas samples for a later laboratory analysis.

5.1 Measurement of trace gases

Table 12. Overview of in-situ observational methods for trace substances (modified from Burrows et al. 2007).

species	instruments/methods
O_3	UV absorption spectrometry, chemiluminesence, electrochemical sonde, mass spectrometry (MS)
O	resonance-fluorescence and optical absorption
H_2O	Lyman-α-resonance-fluorescence, hygrometer, frost point hygrometer, laser photoacoustic spectroscopy (LPAS), tunable diode laser (TDL), cryogenic sampling
OH/HO_2	laser-induced fluorescence (LIF), chemical ionisation mass spectroscopy combined with matrix isolation (CIMS)
H_2O_2	denuder
H_2	cryogenic sampling, residual gas analyzer (RGA)
RO_2	peroxy radical chemical amplifier (PERCA), CIMS
CO_2	infrared CO_2 analyzer, gas chromatography (GC), TDL
CO	vacuum-ultraviolet (VUV) spectrometry, TDL, RGA
CH_4	GC, TDL
VOC, NMHC	proton transfer reaction mass spectrometry (PTR-MS), gas chromatography mass spectrometry (GC-MS)
CH_2O	TDL, denuder with high pressure/performance liquid chromatography (HPLC)
$(CH_3)_2CO$	CIMS
PAN	GC, CIMS
CH_3CN	CIMS
Cl, ClO, BrO	chemical conversion, resonance-fluorescence
ClOOCl, $ClONO_2$	thermal dissociation/chemical conversion, photochemistry, resonance-fluorescence
HBr, HCl, HF	filter analysis with chromatography
NO, NO_y	chemiluminescence with/without gold catalyzer
NO_2	photolysis, LIF
NO_3	cavity ring down spectrometry (CRDS)
N_2O_5	CRDS, LIF
HNO_3	CIMS
HNO_2	TDL
alcyl nitrate	GC with chemiluminescence
SO_2, H_2SO_4, MSA, DMSO	CIMS
OCS	quantum cascade laser
NH_3	CIMS
isotopomers	air sampler with GC and/or MS, TDL
radon	filter with photo-multiplier tube
small ions, electrons	MS, Faraday cup photochemistry
aerosols	impactor, laser-optical particle instrument, forward scattering spectrometer probe (FSSP), growth chamber, two-channel condensation nuclei counter, filter analysis, laser ablation mass spectrometry (AMS)

Table 13. Overview on techniques to detect trace gases.

method/technique	basic principle
photometry/spectrometry	determination of radiation absorption by a gas
fluorescence	gas is stimulated to fluoresce by UV light
flame photometry	gas is stimulated to emit light by a flame
flame ionisation detection (FID)	gas is ionized by combustion, detection of the ions in an electrical field
mass spectrometry (MS)	gas is ionized and then deflected in a magnetic field
gas chromatography (GC)	gas and carrier gas move through a column
liquid chromatography (LC)	gas is dissolved in a liquid, liquid moves through a column
electron capture detection (ECD)	reduction of an electrical current by electron capture in an ionized carrier gas
photoacoustics	absorption of monochromatic light stimulates sound waves
chemoluminescence	absorbing medium is stimulated to emit light by chemical reaction with the trace gas
colorimetry	chemical reaction with the trace gas changes colour of a solution
conductometry	measurement of the electrical conductivity of a solution
titration	quantification by controlled addition of a reagent
long path absorption photometry	dissolution of a gas in a liquid, then determination of radiation absorption

5.1.1 Physical methods

Physical methods to detect trace gases employ a specific physical property of the trace gas to be detected, e.g., its ability to absorb radiation of a given frequency. The trace gas is not changed chemically during the detection.

5.1.1.1 Photometry/spectrometry

Spectrometry is a spectroscopic technique where the wavelength-dependent absorption of electromagnetic radiation by a gas is measured with a photometer. The general measurement principle of a spectrometer can be described by the Beer-Lambert law giving the transmissivity, T:

$$T = I / I_0 = exp(-\alpha C L) \tag{5.1}$$

with I_0 being the emitted radiation from the radiation source, I the received radiation at the detector, L the distance between source and detector, α the absorption cross-section of the detected gas, and C the concentration of the detected gas. Depending on the gas being detected, different wavelengths are used. Ozone, e.g., has an absorption band at 254 nm, thus UV light is needed. Other gases require visible light while greenhouse gases require infrared light for their detection with a spectroscope. For measurements of trace gases within air, the air can either pass through a light path provided to the exterior of a spectrometer (open path method), or it must be filled into or pumped through a cuvette placed in the interior of the spectrometer (closed path method). At the ends of the measurement path of a spectrometer there is a radiation source (usually a lamp) on the one side and a detector (photocell) on the other side. Ideally, the radiation source emits radiation with a selectable frequency-stable wavelength. Either a tunable laser or a broadband light source with filters is utilized. The tunable range for UV light is rather small whereas CO_2-laser light in the infrared spectrum can be tuned from 9 to 11 µm (Colls 2002). Even wider variations are offered by tunable diode lasers where an appropriate semiconducting medium (e.g. a lead-salt) is cooled to the range 70 to 120 K and is forced to emit radiation of a selected wavelength by electrical stimulation. This latter technique is called tunable diode laser absorption spectrometry (TDLAS, Werle et al. 1989). From a mechanical point of view, the open path method is a passive technique whereas the closed path method is an active technique, because the latter requires the forced movement of the air to be analyzed. An overview on various spectroscopic methods is given by Sigrist (1994).

Spectrometers allow for a continuous measurement with a time resolution dependent on the time constant of the detector. With the closed path method, the speed with which the gas can be filled into the cuvette and poured out again is an additional limiting factor. As an example, the cuvette in the CO_2 detecting instrument LiCor 6262 is 0.15 m long, 0.0126 m wide, and 0.0063 m high, thus it has a volume of 0.0119 l. The air flows through this cuvette along the long axis parallel to the light path. A pumping rate of 10 l per minute results in an average residence time of the gas within the cuvette of about 0.07 s (Massman 2004). The tubes leading to the cuvette and the entrance of the cuvette should be coated in such a way that the gas, which is to be detected, is not absorbed or altered by chemical reactions. The pumping of the air through the tubes and the cuvette damps high-frequency variations of the gas concentration. This can lead to an underestimation of fluxes when utilized in an eddy-covariance method (see Ch. 6.3.9).

A recent spectroscopic technique employs a semiconductor-based tunable quantum cascade laser (QCL) which can emit infrared radiation between about 3 µm and 150 µm. Its working principle can be compared to a water fall. The electrons cascade downwards over a sequence of energy levels with equal distance, which have been designed during the growth of the crystal and at each energy step, a photon is emitted. Therefore, QCL can emit up to a factor of thousand more radiation than a diode laser. Due to this efficient process, relatively low efforts for cooling must be made during operation (Wysocki et al. 2004).

Cavity-ring-down-spectrometry (CRDS) is another absorption spectrometric method. It has a about 0.5 m long cuvette ("ring-down-cavity"), through which the gas is pumped, and which is enclosed from both sides by highly effective mirrors. This results in an effective measurement path of about 3 km. The gas in the cavity (which acts as a resonator) is irradiated with a pulsed laser. The diminution of the laser pulses due to reflection losses and absorption by the gas is measured. A CRDS operated at 662 nm permits the detection of NO_3 and N_2O_5 with an accuracy of 0.5 pptv and a time resolution of 5 s. For NO_3, the measurement accuracy is comparable to the one achieved with a DOAS but with a better time resolution. For N_2O_5, no comparable measurement technique exists. NO_3 can be measured at ambient temperature. To detect N_2O_5, the air sample has to be heated to 80 °C, which results in a thermal segmentation of N_2O_5 to NO_3. CRDS then detects the sum of NO_3 and N_2O_5 (Brown et al. 2002).

5.1.1.2 Fluorescence

Some molecules, after they have been excited by the absorption of high-energy radiation, emit a light pulse of a characteristic frequency when returning to their ground state. This luminescence effect is called fluorescence and can be employed for the detection of these molecules. For performing this fluorescence method, a gas sample is filled in a cuvette and exposed to UV radiation. A photocell detects the emitted light which is specific for the trace gases in the sample. This method is especially suited for the detection of SO_2. SO_2 emits radiation with a wavelength between 320 to 380 nm when irradiated with UV light of wavelengths 190 to 230 nm. Water vapour, which suppresses fluorescence, and other gases that could enhance the fluorescence effect, must be eliminated from the gas sample before the measurement (Colls 2002). If laser light is utilized, the method is called laser induced fluorescence (LIF, Kinsey 1977). If a two-dimensional sheet of laser light is produced, the technique is named planar LIF (PLIF, Hanson et al. 1990). With LIF such gases as OH, NO, O_2, C_nH_m, H_2, and H_2O can be detected.

5.1.1.3 Flame photometry

It has been known for a long time that certain substances, e.g. salts, entered into flames give coloured flames. This effect can also be used to quantitatively analyse trace gases. Molecules of the trace gas are excited by the flame and emit light of characteristic frequencies upon return into their ground state. The emission is detected by a photocell and appropriate filters in front of the photocell (Herrmann & Alkemade 1963). The method is used for the determination of sulphur and phosphor concentrations.

5.1.1.4 Flame ionisation

Burnable organic trace gases can be pyrolized into positively charged ions by a hydrogen-air flame. The ions can then be detected in an electrical field built around the flame. The operation of such a flame ionisation detector (FID) is used to determine the total carbon content of atmospheric trace gases. This is because the number of produced positive ions is proportional to the number of carbon atoms. However, CO, CO_2, and HCHO cannot be captured. The detection of ions in a electrical field is a physical method, but the pre-conditioning in the flame is a chemical process which alters the trace substances. FIDs are used to analyze gases which leave the column of a gas chromatograph (Padley 1969).

5.1.1.5 Mass spectrometry

Mass spectrometry is used to determine the weight of ions (de Hoffmann & Stroobant 2007). However, different ions with the same weight cannot be differentiated. The sample gas is first ionised. The ions are then sent through a magnetic field where they are deflected and registered at several counters as a function of the deflecting angle (see Fig. 40). The deflection angle becomes larger with increasing electric charge of the ion while it becomes smaller with increasing mass of the ion. For a statistically sound detection, higher concentrations of the trace substance are required. Therefore, an accumulation step, which collects the trace gas on activated carbon or on other adsorbents, must be performed prior to the operation of the mass spectrometry. For this reason, a mass spectrometer is not suited for a continuous measurement.

There are several methods used to ionize the sample gas. Apart from firing electrons at the sample, chemical methods also exist (Munson & Field 1966). One such method is the proton transfer reaction (PTR) of trace gases with H_3O^+ ions. These ions bond with all VOCs that have a higher proton affinity than hydrogen. The ions, however, do not react with natural constituents of the air. Such a mass spectrometry is called PTR-MS and is suited for an online detection of VOCs. It may also be applied for the determination of gases having concentrations in the ppt-range (Hansel et al. 1999).

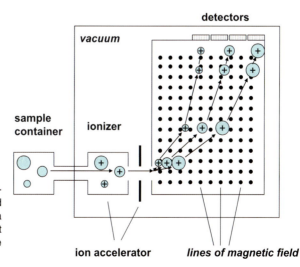

Fig. 40. Schematic of a mass spectrometer. The sample gas is pumped through an ionizer, then deflected in a magnetic field and finally registered at several counters as a function of the deflecting angle.

The efforts for coating and extracting of the above described denuders are quite high (Keuken et al. 1988, Landis et al. 2002). The development of a thermal desorption denuder removes this necessity to regularly coat and extract the tubes (Landis et al. 2002) for certain trace substances, e.g., $HgCl_2$ or $HgBr_2$. At the end of the exposure time, thermal desorption denuders are heated up to between 500 and 700°C and the adsorbed trace substance is liberated to be subsequently analyzed. Thermal denuders allow for an automated operation, therefore manual laboratory work is no longer necessary. Such automated denuders for various gaseous species are reviewed in Slanina & Wyers (1994).

5.1.2.2 Chemoluminescence

Trace gas detection by chemoluminescence is based on reactions of the trace gases with special absorption reactants that excite light emission during these reactions. For example, the blue light emitting dye Coumarin 47 can be employed to detect ozone. The selectivity (see Ch. 2.7) of this method is not perfect because the presence of SO_2 and water vapour may produce some interference. The method was originally developed to detect stratospheric ozone. In the stratosphere, SO_2 and water vapour concentrations are so low that this interference was negligible. The detection limit of the method is 50 ppt; measurement rates of 10 Hz are possible (Güsten et al. 1992).

Ozone may also be detected via chemoluminescence when reacting with nitrous oxide. This technique allows for measurement rates of up to 1 Hz (see e.g. Greenhut et al. 1984). The gas sample is pumped into a reaction vessel where NO is added. About 10 % of the ozone then reacts with the NO to form excited NO_2^*, which returns to the ground state by emitting light of wavelengths between 600 and 3200 nm (the maximum emissions at 1200 nm). This photon emission may be detected with a photo multiplier sensitive for red light. For constant ambient conditions the detected photon emission is proportional to the ozone concentration (Lenschow et al. 1981).

The light emitting reaction between NO and ozone may also be employed inversely to detect nitric oxides. In this case, ozone is added to the gas sample in the reaction vessel. This method allows for the determination of NO and NO_x. NO_2 can subsequently be determined from the difference between NO_x and NO. For the NO_x measurement, the gas sample must pass a converter made, e.g., from hot molybdenum, which reduces the NO_2 into NO before the measurement in the reaction vessel. To determine the NO_2 concentration, NO and NO_x measurements are either performed alternatingly or in two parallel reaction vessels. Further chemiluminescent techniques applied to atmospheric nitrogen compounds are reviewed in Navas et al. (1997) and some practical considerations are also discussed therein. In addition to NO and NO_2, methods for the determination of nitrate, nitrite, nitric acid, nitrous acid and alkyl nitrates are described. Possibilities and limitations of the various procedures are also evaluated in Navas et al. (1997).

5.1.2.3 Colorimetry

Colorimetry is the determination of the colour of a solution with the help of a colorimeter. These instruments measure the absorbance of a specific wavelength of light by the solution. The sample gas is pumped through a tube filled with an appropriate reagent that absorbs the wanted pollutant (SO_2 or NO_x) present in the air to form a complex. This complex is then made to react with other chemicals to form a second, coloured complex. The intensity of the colour is then measured by means of a colorimeter. For example, Saltzman's reagent containing sulanilic acid and acetic acid (or modifications of it) may be used for the detection of nitric oxides (Saltzman 1960, Yanagisawa et al. 1966).

5.1.2.4 Conductometry

Conductometry is based on the measurement of the electrical conductivity of a solution which reacts with the trace gas. The gas sample is pumped through a tube filled with the reagent and the electrical current conducted by the solution is measured. For example an aqueous hydrogen peroxide solution (H_2O_2) may be employed for the detection of SO_2. Sulphuric acid (H_2SO_4) forms when H_2O_2 and SO_2 combine. The quantity of H_2SO_4, and hence the electrical conductivity of the solution is proportional to the amount of SO_2 in the original gas sample.

5.1.2.5 Titration

In a titration method, a reagent, also called the titrant or titrator, is added in small amounts to a solution of the analyte or titrand, whose concentration is unknown. The addition is stopped when an indicator (e.g. the change of colour of the solution) signifies that the endpoint is reached. From the exact amount of the titrant, the unknown amount of the analyte can be determined. For example, SO_2 concentrations in a gas sample may be determined by pumping the gas sample through an aqueous hydrogen peroxide solution (as with the conductometric method described before) and a subsequent titratic determination of the amount of formed sulphuric acid.

5.1.2.6 Long path absorption photometry

The measurement of some trace gases with denuders and subsequent analysis of the adsorbent are difficult due to not fully understood interferences. For one of these gases, HONO, Long path absorption photometers (LOPAP) have been developed recently (Heland et al. 2001). This spectroscopic method is described here together with the chemical methods because it involves a preparatory chemical reaction. HONO is separated in a wet chemical reaction from the air flow by sulfanilamid under formation of diazonium salt. This diazonium salt is converted by further chemical reactions into an azo dye that is then filled into a long Teflon tube. Visible light is focused into the tube and detected at the other end by a spectrometer. The light absorption at 544 nm is proportional to the content of nitrite in the azo dye and thus to the HONO concentration. A detection limit of 1 to 2 pptV can be achieved with an absorption length of 2.5 m. Here, the response time is about 4 min (Kleffmann et al. 2002).

5.1.3 Recommendations for the measurement of trace gases

Many of the above mentioned methods are suited for some specific trace gases only, so that there is a limited choice of the means by which trace gases may be detected. If several methods are available, then possible interferences with other trace gases should be considered when selecting a method.

Continuous gas measurements usually require pumping the gas sample through the instrument. While for inert gases no difficulties are encountered during this pumping, special tubes should be used for reactive gases in order to avoid chemical reactions in the gas on its way to the measurement vessel inside the instrument. The time resolution of such measurements depends on the selected method, the length of the inlet tube, and the flow rate through the instrument. Gas measurement devices must be purified by pumping clean air through them between each run and before calibrations. Calibration is made by pumping special test gases with known concentrations of the trace gas through the instrument.

Methods such as chromatography or mass spectroscopy require sampling at the observation site and a later investigation of the probe in the laboratory. The transport of the samples should be made in containers made from polished stainless steel to avoid any 'aging' of the sample before it is analyzed. The advantage of an laboratory analysis is often a much higher accuracy than can be reached with in-situ instrumentation.

If the trace gas concentration at the measurement site exhibits larger spatial inhomogeneities, measurements made at a single site (by in-situ measurements as well as by sampling for a subsequent analysis) may have a low representativity. In such cases, the employment of path-averaging methods (see Ch. 10.1.1) may be an alternative. A larger number of guidelines for the measurement of trace gases are listed in the Appendix.

5.2 Particle measurements

Solid atmospheric trace constituents (henceforth called particles) are characterized by their physical and chemical properties. The physical properties comprise the size distribution, the structure, the electrical charge, and the solubility of the particles. The chemical properties depend on the chemical composition. Particles may be basically classified into three modes with respect to their formation:

5.2 Particle measurements

- nucleation mode (< 0.1 µm), particles comprising a few molecules of material resulting from atmospheric reactions between gaseous pollutants,
- accumulation mode (0.1–2 µm), particles formed by condensation and agglomeration from the nucleation mode, or produced by industrial activity (e.g. metal fumes and fine combustion products),
- coarse mode (>2 µm), mainly produced mechanically or from sources such as ash from coal combustion.

With respect to their impact on the respiratory system of human beings they may be classified into:
- large particles (> 10 µm), non-inhalable particles, settled in mouth or nose,
- fine particles (< 10 µm: PM_{10}, < 2.5 µm: $PM_{2.5}$), inhalable particles, breathed into the lungs to any extend,
- ultrafine particles (< 1 µm, PM_1), inhalable particles, breathed deeply into the lungs, thus being the most harmful ones.

With regard to deposition it may be differentiated between sedimenting (> 10 µm) and non-sedimenting particles. The determination of all fractions of particles together gives the measure 'total suspended particles' (TSP).

First, we present measurement techniques sensitive to the particle mass, then those sensitive to the particle size. A review on particle measurement methods is given in Colls (2002).

5.2.1 Determination of the particle mass

A variety of methods is available for the determination of particle mass (Tab. 14). The first method is specially designed for sedimenting particles, the other methods can be applied to all particle size fractions.

Table 14. Overview of methods to determine the particle mass.

method/technique	basic principle
deposit gauges	deposition into an open bottle
sampler	air stream is sucked through a filter
nephelometer	measurement of scattered radiation
ethalometer	measurement of light absorption on a filter
β-radiation absorption	β-radiation absorption on a filter
pressure drop at filters	measurement of pressure drop across a filter
resonance methods	resonance properties of a carrier (crystal) changed by deposited particles

5.2.1.1 Deposit gauges

Deposit gauges are simple and inexpensive collecting instruments especially suited for long-term monitoring. A glass bottle with a funnel fixed to its top is mounted on a pole roughly two metres above the ground. It is exposed for some days or even a month and collects particles from dry and wet deposition. After the exposure, the glass is dried by evaporating residual water and the remaining dry substance is weighed. The mean particle deposition flux is determined by relating the dry particle mass to the funnel size and the exposure time. In Germany this technique is called the 'Bergerhoff method'.

5.2.1.2 Sampling on filters

The next simple method to deposit gauges is to collect larger amounts of particles for a subsequent analysis on filter papers. For this purpose, a well-defined amount of polluted air is drawn through a piece of filter paper with a specified porosity. The particle mass is determined from the weight difference of the filter before and after the exposure. The two weighing processes are to be made by equal ambient conditions (relative moisture and temperature). The filters should be kept at low temperatures to avoid evaporation of organic substances before weighing them. The filter material should not react chemically with atmospheric gases and particles. Filters made from glass fibres can absorb sulphuric acid, those made from quartz fibres and cellulose can absorb water vapour (Colls 2002).

5.2.1.3 Nephelometers

Nephelometers register the intensity of light scattered by particles. They are also called aspirated smoke detectors, if they are used for the detection of particles from combustion processes. The air sample is delivered into a 2 m long and 15 cm wide measurement vessel through which a light beam is sent. The light intensity may be measured at right angles to the light beam or from the end of the vessel opposite to the light source. In the former case scattering, in the latter case the extinction is determined. If light of different wave lengths is used, differences in the observed scattering can give information about the size distribution of the particles (Colls 2002). Nephelometers are suited for continuous operation. They are also employed for visibility monitoring. Recent improvements of nephelometers are described by Varma et al. (2003).

5.2.1.4 Aethalometers

Aethalometers measure the light attenuation by accumulated particles. The total mass of black carbon (soot) on a filter can be determined after a defined volume of air had been drawn through the filter. After the exposure of the filter, the light transmission ability of the filter is determined by an aethalometer. A broad-band light source is mounted on one side of the filter and light detectors on the other side. The

5.2 Particle measurements

measured light attenuation is proportional to the deposited amount of black carbon on the filter (Colls 2002). Arnott et al. (2005) compare aethalometer black carbon measurements to those with other instruments.

5.2.1.5 β radiation absorption

Beta attenuation monitors (BAM, Fig. 41) analyse the attenuation of β radiation by accumulated particles. The total mass of particles on a filter can be detected after a defined volume of air has been drawn through the filter. After a selectable sampling period of 30 minutes to 24 hours, the filter is exposed to radiation emanated from a β radiation source (e.g. krypton 85) and the received radiation is investigated by a radiation detector from the opposite side of the filter. The attenuation of β radiation by the filter is proportional to the accumulated particle mass on the filter. The particle size distribution and their chemical composition have neligible influence on the measurement. The choice of size-specific inlets allows for a size preselection of the particles, e.g., for $PM_{2.5}$ or PM_{10}. An accuracy of 1 µg/m³ can be achieved for sampling periods of 24 hours. The measurement can be automated by utilization of long filter tapes which are forwarded by an electrical motor after the end of each sampling period (Fig. 42). The exposed spot of the tape is transported to the detection section of the instrument. This allows for long-term unattended operation of BAMs. For a comparison of BAM with other particle concentration monitors see Chung et al. (2001).

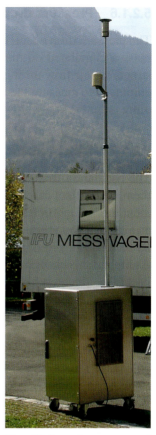

Fig. 41. Aerosol measurement by beta absorption. The inlet is mounted on top of the pole. A schematic of the measurement principle is shown in Fig. 42.

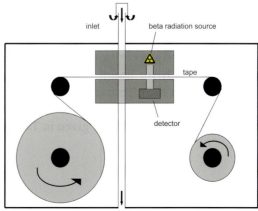

Fig. 42. Schematic of automated PM_{10} monitoring by beta radiation attenuation. After the end of each accumulation period the tape is moved a bit so that the exposed spot of the tape is then located between the radiation source and the detector. The instrument is shown in Fig. 41.

all stages of the impactor, highly sensitive electrometers register the electric charge (currents of down to some 10^{-15} Ampère can be detected) of the particles collected at the disks. Particle sizes from about 30 nm to some µm can be recorded in about 12 stages with this type of impactor.

5.2.2.2 Differential mobility particle sizers

Differential mobility sizers (DMS) or differential mobility analysers (DMA) analyse the size-dependent mobility of charged particles in a electrical field (Fig. 44). This technique requires particles to be ionised at the inlet of the instrument. In instruments such as a scanning mobility particle sizer (SMPS) or an electrical aerosol analyser (EAA), krypton 85 is used for this purpose. Thereafter, the charged particles are drawn through an electrical field. The deflection of the particles in the field is larger for smaller particles. Several detectors mounted side by side at the end of the measurement path count the impacting particles with respect to the gained deflection angle. A single measurement takes about two to three minutes and differentiates particles between 10 and 500 nm (Colls 2002). Recently, fast mobility particle sizers (FMPS) have been developed which allow for the analysis of up to 10 particle size spectra per second. This even permits the employment of particle sizers in eddy-covariance methods for particle flux measurements (Horn 2005, see also Ch. 6.3.9). The observation of larger particle size ranges is possible by twin differential mobility particle sizers (TDMPS) which are based on two simultaneously operated DMA with different size ranges (Birmili et al. 1999).

5.2.2.3 Optical particle sizers

Light scattering from particles depends on the particle size and can therefore be employed to measure particle size distribution. In an optical particle sizer, a small measurement volume of a few mm³ is irradiated by a laser beam. The airflow through this

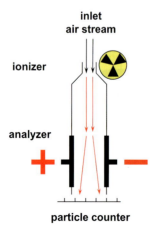

Fig. 44. Schematic of a differential mobility particle sizer (DMPS). Particles in the incoming flow are ionized by a krypton 85 source. The electrical field deflects the charged particles. The deflection angle is inversely proportional to the particle mass.

volume is tuned in such a manner that only one particle at a time can be expected to be within the volume. The intensity of the scattered light is detected by a photo cell and after a sampling time of a few minutes the light intensity distribution can be converted into a particle size distribution. The lower end of the detectable spectrum is at about 300 nm. Calibration of an optical particle sizer must be made with a sample aerosol whose shape and structure is close to the aerosol which is to be detected, since the scattered light is sensitive to these parameters (Colls 2002).

5.2.2.4 Aerodynamic particle sizers

Particles in an accelerating air stream lag the flow due to their inertia. The velocity difference between the air and the particles increases with increasing particle size and can be employed for particle size measurements (Baron 1986). In an aerodynamic particle sizer (APS), an air stream is accelerated through a nozzle and the speed of the particles transported by this stream is measured when the particles pass two partly overlapping parallel laser beams. Like the optical particle sizer described before, the aerodynamic particle sizer is a single-particle-counting instrument. The light scattered by a particle flying through the two laser beams produces a two-crest signal. The distance between the two crests in the signal is inversely proportional to the particle speed and directly proportional to the particle radius. The instrument is suited for particles between 0.5 and 20 µm and measurements with an accuracy of 0.02 to 0.03 µm. In contrast to the optical particle sizer described before, the measurement does not depend on the surface structure of the particles.

Fluorescence aerodynamical particle sizers (FLAPS) have been developed which additionally send the particles to a UV beam of 355 nm and then record any resulting fluorescence with a photomultiplier (Hairston et al. 1997). FLAPS are used to distinguish between aerosol of biological (which may exhibit fluorescence) and non-biological origin.

APS methods determine the so-called aerodynamic radius of the particles which corresponds to the radius of a spherical particle exhibiting the same flow resistance. This aerodynamic radius has direct relevance in studies on the inhalability of particles into the respiratory system and thus on health impacts.

5.2.2.5 Condensation particle counters

Particles can act as condensation nuclei and trigger condensation in saturated air. This property is utilized in condensation particle counters (CPC) or condensation nucleus counters (CNC) (Argawal & Sem 1980). Because this property does not depend on the particle size, a size classifying stage (e.g., a DMA, see above) has to be operated before the particles of a selected size are allowed to enter the CPC. Inside the CPC, the particles enter a measurement vessel filled with air which is oversaturated with water or alcohol (n-butanol, iso-propanol) vapour, and this triggers condensation. The emerging droplets are detected by light scattering. The intensity of the detected scattered light is proportional to the number of particles which have entered the measurement vessel because the drop size does not depend on the size of the condensation nucleus. The lower end of particle size distribution detectable by a CPC is at about 3 nm (Horn 2005). If two many small particles enter a CPC, the oversaturation in the measurement chamber is reduced to fast and not all particles are able to trigger condensation. This leads to an underestimation of the number of very small particles. A comparison between 26 CPCs is reported in Wiedensohler et al. (1997).

5.2.2.6 Electrical aerosol detectors

The electrical charge of unipolarly charged aerosol particles is proportional to the 1.13[th] power of the particle diameter (Lawless 1996, Horn 2005) so that a measurement of the electrical charge of particles tells something about their size. In an electrical aerosol detector (EAD), the incoming flow is splitted and one part of the entering aerosol particles is ionized when flowing past a corona needle. In a mixing chamber these ionized particles transfer their charge by diffusion charging to the rest of the aerosol. Finally an electrometer detects the charge of the particles.

5.2.3 Measurement of the chemical composition of particles

A variety of special methods are available for the determination of the chemical composition of particles (see Tab. 16 and in a much more detailed form in Appendix B in McMurray et al. 2004). Some of the above mentioned analysis techniques for gaseous species can be employed if the chemical species in the particles can be converted quantitatively into specific gases. For example, $SO_4^=$ ions, after gaseous SO_2 has been removed by piping the sample trough a denuder (see Ch. 5.1.2.1), can be detected by flame photometry (see Ch. 5.1.1.3) because at a temperature of around 1000 °C, $SO_4^=$ is converted into SO_2 (McMurray et al. 2004).

Table 16. Overview of methods for the determination of the chemical composition of particles.

method/technique	basic principle
conversion of carbon to CO_2	measurement of CO_2
atomic absorption spectroscopy	heating and subsequent absorption measurement
emission spectroscopy	exciting by strong heating and subsequent emission measurement
x-ray fluorescence and proton capture spectroscopy	element specific radiation emission after excitement by proton or x-rays

5.2.3.1 Thermal evolution of carbon

The carbonaceous component of atmospheric particles consists of black carbon (BC), which is often described as elemental carbon (EC) or soot, and organic carbon (OC), which is composed of hundreds and even thousands of individual organic compounds. Both fractions together are designated as total carbon (TC). Measurement of particulate carbon can be performed by automatic thermal evolution of CO_2 at 340 °C (adjustable) for OC and 750 °C for TC. The carbon collected on a ceramic impactor plate is oxidised at the aforementioned temperatures after the sample collection is complete. A CO_2 monitor then measures the amount of carbon released as result of the oxidation (McMurray et al. 2004).

5.2.3.2 Atomic absorption spectroscopy

Particles can be decomposed into gases and atoms by hot flames (2000 to 3000 °C) or over electrically heated surfaces (4000 to 6000 °C), and subsequently analyzed in a spectroscope. The method, called atomic absorption spectroscopy (AAS, Ranweiler & Moyers 1974), is suited for the detection of heavy metals in particles.

5.2.3.3 Atomic emission spectroscopy

The application of particle decomposition for the aforementioned atomic absorption spectrometry may produce excited atoms. These atoms will return to their ground state by emitting radiation. If the detection method for the analysis of the chemical composition of particles is based on the analysis of this emitted radiation, then it is called atomic emission spectroscopy (AES). Like the AAS (Ch. 5.2.3.2), this method is suited for the detection of metals in particles. A special AES method is the inductively-coupled plasma AES (ICP-AES, Fassel & Knisely 1974) where the sample is brought into an argon plasma. At temperatures of more than 6000 °C, the valence electrons are excited to a higher level from which they return by emitting radiation. The frequency of the emitted radiation is specific for the chemical species. The intensity is proportional to the concentration of the species in the sample.

5.2.3.4 X-ray fluorescence and proton-induced x-ray emission

Related to the atomic emission spectroscopy are the x-ray fluorescent spectroscopy (XRF, Török & van Grieken 1994) and the proton-induced x-ray emission (PIXE, Johansson et al. 1975). In these methods, the sample is irradiated with x-rays or protons which remove electrons from the inner shells of atoms. Upon return to the inner shell, these electrons emit x-rays that can be analyzed. The frequency of the x-rays is species-specific, their intensity is proportional to the concentration of the species in the sample.

5.2.4 Measuring the particle structure

The efficiency of a particle as a condensation nucleus or the health impact of an inhaled particle does not only depend on its mass, size, or chemical composition but also on its surface structure (Maynard 2007). Furthermore, the particle structure may give hints as to the origin of the particles, e.g., if they are identified as pollen or as specific minerals (Bernabé et al. 2005). Differentiation between primary and secondary particles and between coagulates and agglomerates are possible (Poppitz & Heidenreich 2005). The particle surface structure may be investigated by observations with an optical microscope or a scanning electron microscope (SEM, see, e.g., Fruhstorfer & Niessner 1994).

5.2.5 Saltiphon

Sands may be moved by stronger winds, a transportation process called saltation (Shao & Raupach 1992). A saltiphon has been developed to detect saltation. It consists of an acoustically sensitive diaphragm with a 10 cm diameter mounted into a steel tube. A saltiphon counts the number of impacts of sand grains as function of time. A wind vane helps to keep the tube parallel to the wind direction (Spaan & van den Abeele 1991). This detection principle has similarities with a microphone.

5.2.6 Recommendations for particle measurements

Often, particle number and size measurements may be performed continuously. For this purpose, the sample air is piped through tubes into the analysis device. The inlet should be as straight as possible and without any obstacles to keep premature deposition of particles as low as possible. If the particles are collected on a filter tape, then collection of a new sample and analysis of the previous sample can often be performed in parallel. Mostly, the step time for forwarding the tape is dominated by the required period for the collection of a sufficient amount of material for the subsequent analysis. The flow rate of the sample gas through the analyser has to be known with high accuracy, because any uncertainty in the flow rate adds to the uncertainties of the detection method.

Some analysis methods can only be operated in a laboratory such as, e.g., electron microscopy and many methods for the analysis of the chemical composition. Mostly, sampling on filters is sufficient for a later laboratory analysis. This analysis should be made shortly after the sample collection to keep changes of the sample due to evaporation, hygroscopic uptake of water, and chemical reactions as low as possible.

The appendix lists a larger number of technical guidelines for particle analysis.

5.3 Olfactometry

Odours are caused by the presence of certain aromatic molecules in the air. The number of these aromatic molecules is often so low that the previously described physical and chemical methods as well as long-path remote sensing methods (see Ch. 7 below) are insufficient to detect them. Furthermore, the exact chemical composition of many odours is not known. Therefore, odour measurements are often made with the help of specially trained persons (assessors). In such an assessment, the amount of an odourous substance is determined as the multiple of the minimal amount which can be recognised by the assessor. The analysis is made with the help of an olfactometer, an instrument that provides repeatable dilutions of the air sample. Automatic detection systems have been developed recently for those odours whose exact chemical composition is known. These so-called "artificial noses" or "electronic noses" (see Nimmermark 2001 for a detailed review) can, e.g., utilize one of the two following methods.

The resistance method uses semiconducting metal oxides (MOX) as sensors for the detection of gaseous parts of the air. The electrical resistance of these metal oxides changes when a surface reaction with the sought aromatic gas molecules takes place. The selectivity of the MOX sensors is influenced by the choice of the metal oxide and its doping with selected impurities (Zudock 1998).

The resonance method analyzes the resonance frequency of so-called surface acoustic wave elements (SAW elements) that is proportional to their mass. SAW elements are an inexpensive mass product from the electronic industry. The surface of these elements has been coated to make them sensitive for selected gases. Their mass increases and their resonance frequency changes, if such gases are deposited from a sample air flow (FZK 2000).

Biosensors for olfactometry have also been developed. Such a biological sensor system can be sensitive to different odorants. The system consists of a sensor in which isolated olfactory receptor proteins (ORPs), e.g., from bullfrogs, were coated onto the surface of a piezoelectric electrode, similar to the mechanism of human olfaction. The electrode serves as a signal transducer. Results indicate rapid (about 400 s), reversible, and longterm (up to 3 months) stable responses to different volatile compounds such as n-caproic acid, isoamyl acetate, n-decyl alcohol, β-ionone, linalool, and ethyl caporate. The sensitivity of the sensor ranges from 10^{-6} to 10^{-7} g, which is fully correlated with the olfactory threshold values of human noses (Wu 1999).

Technical rules for olfactometry may be found in the VDI guidelines VDI 3881, VDI 3882, and VDI 3940 (see Appendix) and in the European standard CEN 13725.

6 In-situ flux measurements

All methods presented so far in Chapters 3 to 5 have been designed to determine either a state variable of the atmosphere, the mass or concentration of moisture, or an atmospheric trace substance. These quantities alone are not sufficient to describe the geobiochemical cycles within the Earth's system and to close the energy and substance budgets in the atmosphere. For this closure of these budgets and cycles, we also need the mass, energy, and momentum fluxes between the different compartments of the Earth's system (atmosphere, biosphere, soil, etc.) and the incoming and outgoing radiation fluxes. We start with the radiation fluxes.

6.1 Measuring radiation

Table 17 lists terms necessary to characterize radiation and radiation fluxes. For example, the solar energy density which amounts in Central Europe to about 9 kWh m^{-2} on a sunny summer day, is a radiation energy density. The solar constant of 1369 W m^{-2} is a radiation flux density.

In observing radiation, a distinction must be made between shortwave and longwave radiation. Incoming solar radiation is shortwave radiation, outgoing terrestrial or thermal radiation is longwave radiation. The sum from both is called total radiation. The difference of downwelling and upwelling radiation is called net radiation. Shortwave radiation may be differentiated between direct radiation from the sun and diffuse radiation (i.e. scattered sunlight). The sum from both is called global radiation.

Radiation is detected with radiometers. Radiometers for visible light are called photometers. Spectral radiometers see a limited wavelength range only. Spectrometers record the wavelength-dependent irradiance. Radiation measurements can be made either by the photovoltaic method or by the thermometric method (see Tab. 18). The photovoltaic method is based on the photoelectric effect. Incoming radiation on a semiconductor triggers an electric current proportional to the radiation flux density. This is the same principle by which solar cells convert solar energy into electricity. The main disadvantage of the photovoltaic method is that it depends strongly on the wavelength of the incoming radiation. Therefore, the thermometric method, which is based on heating of a body with a black surface due to radiation absorption, is most frequently used. This method is insensitive to the wavelength of the incoming radiation. The temperature increase of the sensor is measured with resistance thermometers or thermoelectric elements (see Ch. 3.1). Absolute instruments measure the temperature increase of the sensor, relative instruments measure the temperature difference between two sensors, one with a white and the other with a black surface.

6.1 Measuring radiation

Table 17. Terms describing radiation and physical units in which they are recorded (see also Tab. 1 and 2).

quantity	description	physical unit
energy	amount of energy transferred by radiation (time integrated radiation flux)	1 J = 1 W s, larger unit: 1 kWh = 3,600,000 W s
energy density	radiation energy flowing through a unit area (time integrated radiation flux density)	1 J m^{-2} = 1 W s m^{-2}, larger unit: 1 kWh m^{-2} = 3,600,000 W s m^{-2}
flux (power)	received or emitted radiation energy per unit time	1 J s^{-1} = 1 W
flux density	received or emitted radiation energy per unit time and area	1 J s^{-1} m^{-2} = W m^{-2} [non-SI unit: 1 W m^{-2} = 1.433 · 10^{-3} cal cm^{-2} min^{-1} = 1.433 · 10^{-3} langley min^{-1}]
intensity	flux (radiative power) per unit angle	1 W sr^{-1}
radiance	intensity per unit area	1 W sr^{-1} m^{-2}
irradiance	incoming radiation flux density	1 J s^{-1} m^{-2} = W m^{-2}
spectral irradiance	incoming radiation flux density per unit wavelength interval	1 J s^{-1} m^{-2} nm^{-1} = W m^{-2} nm^{-1}
emittance	outgoing radiation flux density	1 J s^{-1} m^{-2} = W m^{-2}
spectral emittance	outgoing radiation flux density per unit wavelength interval	1 J s^{-1} m^{-2} nm^{-1} = W m^{-2} nm^{-1}

Table 18. Overview on radiometer and other radiation measurement instruments.

instrument	quantity	method
actinometer	direct solar irradiance/shortwave irradiance	determining the amount of heating
pyrheliometer	direct solar irradiance	determining the amount of heating
pyranometer	shortwave irradiance	determining the amount of heating
photometer	shortwave irradiance	counting of photons
pyrgeometer	longwave irradiance	determining the amount of heating
pyrradiometer	short- and longwave irradiance	determining the amount of heating
frigorimeter	longwave emission	determining heating energy to counterbalance the cooling
katathermometer	longwave emission	determining the speed of a temperature decrease
sunshine recorder	sunshine duration	length of a burned trace

6.1.1 Measuring direct solar radiation

Actinometers

Actinometers measure the direct solar radiation or other shortwave irradiance (actinic means "caused by rays" in Greek) via the warming of a blackened sensor in comparison to a similar sensor not exposed to radiation. They are relative instruments and need calibration.

Chemical actinometers measure the number of photons via the production of a chemical compound. For example, a benzophenone-benzhydrol system can be utilized for photon detection (Rosenthal & Bercovici 1976).

Pyrheliometers

Pyrheliometers (pyro means fire or heat in Greek, and helios is the Greek god of sun) measure the absolute direct solar irradiance. They are constructed so that only the sunlight but no scattered radiation can reach the sensor. The sensor consists of two blackened stripes. One of the stripes is exposed to the sunlight, the other is heated electrically and shielded from the sunlight. The amount of electrical energy necessary to keep the second stripe at the same temperature as the first one is proportional to the energy of the sunlight. The accuracy of a pyrheliometer is about 5 % (Strangeways 2003).

6.1.2 Measuring shortwave irradiance

Pyranometers

Pyranometers measure both direct solar irradiance and diffuse shortwave irradiance which come from the upper hemisphere over the instrument by determining the temperature difference between a white and a black surface element (or between a black surface element and the air) (Fig. 46). A white surface reflects the incoming shortwave radiation but not incoming longwave radiation. A black surface on the other

Fig. 46. Pyranometer for the measurement of shortwave radiation. A glass dome shields the sensitive black and white sensor surfaces from disturbing environmental influences (mainly wind and precipitation).

6.1 Measuring radiation

Fig. 47. Thermopile pyranometer for the measurement of shortwave radiation. Under the double glass dome the black coating of the thermopile sensor can be seen.

Fig. 48. Measurement of the diffuse shortwave radiation. The ring shields the instrument from direct sunlight (Photo: Helmut Mayer).

hand absorbs both shortwave and longwave radiation. Therefore the difference in heating of the two surface elements is from absorption of shortwave radiation. Other pyranometers use black coated thermopiles as sensors (Fig. 47). Pyranometers are shielded against precipitation and cooling by the wind with a glass dome. Such a glass dome also filters out longwave radiation above 3 µm. The time constant of pyranometers is in the order of 5 to 20 s, their accuracy after diligent calibration is about 5 % (Strangeways 2003). Often, these instruments have a double glass dome (Fig. 47) to avoid that the thermal radiation from the outer dome, which warms up in the sunlight, influences the measurement. The pure diffuse shortwave irradiance can be recorded by shadowing the instrument from direct sunlight with a moveable disk or a fix-mounted stripe (Fig. 48).

Albedometers

If a pyranometer is mounted upside down, so that it receives the radiation from below, then it can record the reflected shortwave radiation from the ground. The ratio between incoming and reflected shortwave radiation is called albedo of the ground.

Fig. 49. Radiation balance measurement. Two identical sensors look to the upper and the lower half space (Photo: Helmut Mayer).

A pair of two pyranometers, one looking upward and one looking downward, is also called an Albedometer (Fig. 49).

Photometers and other spectrally selective radiometers
Radiometers that are sensitive to the same light spectrum as the human eye are called photometers. Sun photometers orientated directly to the sun are special instruments for measuring the optical depth of the atmosphere, a variable depending on the presence of light scattering objects (aerosols, gas molecules) in the atmosphere. Compact light-emitting diode (LED) sun photometers, in which a LED is used as a spectrally selective photodetector as well as a nonlinear feedback element in the operational amplifier, have been developed. The output voltage is proportional to the logarithm of the incident solar intensity. This permits the direct measurement of atmospheric optical depths in selected spectral bands (Acharya et al. 1995).

Other radiometers have been designed to detect the intensity of direct and diffuse UV radiation. There are broad-band radiometers that are sensitive to the whole range of ultraviolet radiation and there are narrow-band spectrophotometers which selectively detect the narrow band of UV-A or UV-B radiation (one of these being the Brewer spectrophotometer that is usually used to determine the optical depth due to the presence of ozone in the atmosphere and thus delivers the ozone column density, see, e.g., Carvalho & Henriques (2000)). The accuracy of these instruments is in the order of 2 to 7 % (Strangeways 2003). Again, other radiometers are sensitive to the photosynthetically active radiation between 400 and 700 nm (PAR) by employing special filters. Ross & Sulev (2000) discuss the exact definition of PAR, the instruments to measure it and possible error sources with this measurement.

The development of the multifilter rotating shadowband radiometer (MFRSR, Harrison et al. 1994) has been the attempt to combine several of these specialized radiometers into one single instrument. This MFRSR consists of a diffuser that directs the incoming shortwave radiation from the upper half space onto a series of photodetectors. One detector registers the whole incoming radiation, while the others (interference-filter photodiodes) record incoming radiation in narrow 10 nm wide frequency ranges (e.g., around 415, 500, 615, 673, 870, and 940 nm). At the outside of the instrument, a metal band rotates to prevent at regular intervals the direct solar light entering the instrument. During these periods of eclipse, only the diffuse radiation is measured by the instrument. Finally, the direct solar radiation is determined from the difference between the total short wave radiation and the diffuse radiation.

6.1.3 Measuring longwave irradiance

An instrument designed to detect only the longwave irradiance is called a pyrgeometer. It is based on the detection of the warming of a white surface element which reflects shortwave radiation but absorbs longwave radiation. The longwave radiation budget is measured by net pyrgeometers. These instruments have domes made from transparent materials permitting only longwave radiation to pass. Examples of such materials are polyethylene or lupolen. Other pyrgeometers use thermopiles with black windows which are transparent for longwave radiation between 4.5 µm and 50 µm (Fig. 50).

6.1.4 Measuring the total irradiance

Measuring both components, the shortwave and the longwave radiation together is achieved by substituting the glass dome of a pyranometer (see Ch. 6.1.2 above) with a dome made from polyethylene which is transparent to both shortwave and longwave radiation. An instrument recording both components of radiation from the upper half space is called pyrradiometer. Arranging two such instruments, one looking upward and one looking downward results in a net radiometer recording the complete radiation balance. A net radiometer can alternatively be realised from a combination of a net pyrgeometer and an albedometer. The achieveable accuracy is comparable to those of pyranometers.

Fig. 50. Pyrgeometer measuring incoming longwave radiation. The black silicon window with a solar blind filter coating has a transmittance between 4.5 µm and 50 µm that eliminates solar shortwave radiation (Photo: Helmut Mayer).

6.1.5 Measuring chill

Attempts have been made to objectively capture human sensitivity to warmth or cooling. Feeling of warmth or cooling (the latter is commonly called chill) is not only due to radiation fluxes but also due to turbulent heat and moisture fluxes (for the latter see Ch. 6.3). Chill due to the combined influences of air temperature, wind, precipitation, and radiation can be measured with a frigorimeter (Siple & Passel 1945). Such instruments try to simulate the heat flux per unit area and unit time emitted from a human body with a skin temperature of 36.5 °C. A frigorimeter consists of a blackened copper sphere with a diameter of 7.5 cm exposed to the ambient air. The sphere is heated electrically and the current necessary to keep it at 36.6 °C is proportional to the heat loss of the sphere.

An alternative method for the determination of chill effects is the use of the kata thermometer developed by Leonard Erskine Hill (1866–1952) in 1923 (Hill 1923). For this method, an alcohol thermometer is heated to 100 °F (roughly 37 °C, the temperature of a human body), and then either with the bulb wiped dry or with a wet muslin jacket wrapped around it, the time it takes to cool down to 95 °F is determined. This time duration is inversely proportional to the chill. The kata thermometer is essentially the wet bulb thermometer from a psychrometer (see Ch. 3.2.2).

6.1.6 Sunshine recorder

The duration of sunshine can most easily be determined with a Campbell Stokes sunshine recorder or sunshine autograph by the burn method (Fig. 51). This device consists of a massive glass sphere with a diameter of 96 mm mounted concentrically in a bowl such that the sun's rays are focused to a card fixed at the inside of the bowl. The refraction index of the glass is about 1.51. Sunshine leaves a burned trace on the card whose length is determined manually afterwards. The time resolution is less than one tenth of an hour. The instrument tends to report a too long sunshine duration. In the case of broken clouds up to twice the correct value can be produced. Annual means can still be in error by about 10 % (Strangeways 2003).

Fig. 51. Campbell Stokes sunshine recorder (sunshine autograph).

Automated data recording today employs pyrheliometers that measure the duration for which the shortwave radiation from the sun is above the threshold of 120 W/m². Likewise, the difference between global and diffuse radiation recorded by pyranometers can be evaluated. Sunshine is assumed to be present as long as this difference, divided by the cosine of the zenith angle of the sun, is above 120 W/m².

Records from these instruments differ up to about 20 %. In former times the Campbell Stokes sunshine recorder was used as reference standard, but now Pyrheliometer are recommended (WMO 2006).

6.1.7 Recommendations for radiation measurements

The success of radiation measurements depends decisively on the choice of the observation site. The instruments have to be positioned in such a way that no obstacles can affect the measurements. Obstacles that rise less than 3° above the horizon can be neglected. Most ideal are sites on top of high-rise buildings.

Instruments observing the lower half space are typically mounted to booms on a mast about 2 m above the ground. The surface below the instrument should be homogeneous and representative of the surroundings. About 90 % of the received signal comes from a half opening angle of 65°. This means that the data obtained from an instrument mounted at 2 m above ground is dominated by the surface properties of a circle underneath with a diameter of 8.8 m (i.e. an area of about 60 m²).

Apart from calibrations, radiation instruments need regular maintenance. Their glass or Lupolen domes have to be wiped regularly. Dust and rain drops modify the measurement results. Plastic domes age and are sometimes subject to attacks by birds. To avoid dew and rime, a heating ring can be mounted to the base of the dome. Longwave thermal radiation from the dome has to be considered in the calibration. Measurements without domes cannot be recommended because wind and precipitation will massively influence the behaviour of the sensors.

Technical rules for radiation measurements can be found in the VDI guideline 3786 part 5 (see Appendix) and in Chapter 7 of WMO (2006). The measurement of sunshine duration is covered in Chapter 8 of WMO (2006).

6.2 Visual range

The visual range or visibility is limited by Rayleigh scattering with air molecules, absorption at NO_2 molecules, and Mie scattering and absorption at aerosols. The maximum possible visual range in an atmosphere without any aerosols and NO_2 molecules amounts to 334 km at a light wavelength of 550 nm. A NO_2 concentration of 250 ppb in an aerosol-free atmosphere lowers this range down to 63.6 km (Jacobson 2002).

A distinction is made between meteorological visual range (daytime visual range) and fire visual range (night-time visual range). Meteorological visual range is defined as that range in which a dark body can still be recognized against its background (2 % contrast). Fire visual range can be considerably larger than the meteo-

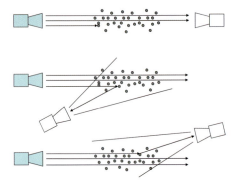

Fig. 52. Schematic of transmissometers (top), backward scatterometers (middle) and forward scatterometers (below) for the measurement of the visual range. The arrows indicate the emitted light beams. Full symbols: emitter, open symbols: receiver.

rological visual range. This is the range from which a bright light source can still be detected. These measures depend on the abilities of observers. Therefore, in 1971, WMO (1990) introduced the definition of the meteorological optical range (MOR). MOR is the path length along which a focused light beam having a colour temperature of 2700 K is attenuated to 5 % of its original intensity.

The visual range can be monitored with transmissometers and with scatterometers (Fig. 52). Nephelometers described in Chapter 5.2.1.3 are special scatterometers. A transmissometer (Douglas and Young 1945) sends a light beam to a separate receiver (bistatic method) or sends it to a mirror which reflects the beam to a receiver integrated within the device (monostatic method). The attenuation of the light beam during its travel is proportional to the amount of aerosol particles and small water droplets along the light path and thus proportional to the visual range. Scatterometers emit a light beam and receivers mounted next to the path of the light beam record the intensity of scattered light. Some methods analyse forward scattering, other methods analyse backward scattering (see Fig. 52). The intensity of the scattered light is proportional to the aerosol particle concentration and thus, again, proportional to the visual range. Today, visual range monitoring instruments are found at nearly all airports and frequently close to motorways to detect fog and mist dangerous to traffic.

Technical rules for determining the visual range with LIDAR methods (see also Ch. 7.2.5) are documented in the VDI guideline 3786 part 15 (see Appendix). An ISO guideline is currently under development. Chapter 9 in WMO (2006) gives further details. Multiband transmissometers (MSRT transmissometers) are used to analyze aerosol characteristics (see, e.g., de Jong et al. 2007).

6.3 Micrometeorological flux measurements

The necessity of flux measurements has already been addressed briefly in Chapter 2.1.5. A review of the major micrometeorological techniques with respect to the measurement of vertical turbulent trace substance fluxes is presented in Lenschow

6.3 Micrometeorological flux measurements

(1995). Techniques focusing mainly on turbulent vertical energy fluxes (sensible and latent heat) are extensively discussed in Foken (2008). The theoretical basis of flux monitoring techniques, especially those for CO_2 fluxes is critically outlined in Lee et al. (2004). A very recent overview of approaches to determine the trace substance exchange between soil and atmosphere with an explicit evaluation of the different methods has been published by Denmead (2008).

Here, we start with the presentation of methods (see also Tab. 19) that do not require concentration measurements with a high time resolution (Ch. 6.3.1 to 6.3.6).

Table 19. Overview on flux measurements.

method	measured quantities	principle
cuvettes	trace substance concentration after accumulation	increase of concentration is proportional to the flux
surface chambers	trace substance concentration after accumulation	increase of concentration is proportional to the flux
mass balance method	trace substance concentrations, wind speeds	integration of budget equations yields fluxes
inferential method	mean trace substance concentration in one height, thermal stratification	flux equals concentration times deposition velocity
gradient method	mean trace substance concentration in two heights, thermal stratification	flux equals concentration gradient times exchange coefficient
Bowen-ratio method	mean trace substance concentration in two heights, net radiation, surface heat flux	flux of trace substance is assumed to be proportional to a known heat or moisture flux
flux variance method	variances of trace substance concentrations	flux is assumed to be proportional to the variance
dissipation method	variances and spectra of trace substance concentrations	integration of budget equations for turbulent kinetic energy and the variances of the trace substance concentrations
eddy-covariance method	simultaneous measurement of wind speed and trace substance fluctuations	flux directly determined from the covariance of concentration and wind speed fluctuations
eddy-accumulation method	accumulation of trace substances in two containers depending on the sign of the vertical wind speed fluctuation	flux is assumed to be proportional to the concentration difference in the two containers
disjunct eddy-covariance method	short-time wind and trace substance concentrations at short time intervals	flux is computed from the covariances of the fluctuations of wind speed and concentration
evaporation measurements (Ch. 6.4)	amount of evaporated water	weighing or water level readings
inverse methods (Ch. 6.6)	trace substance concentrations downwind of the source	flux is computed by inverse dispersion modelling

Chapters 6.3.7 and 6.3.8 will mention two methods which do require high resolution trace substance measurements. Chapter 6.3.9 will present the only direct and unfortunately the most laborious method, the eddy covariance method. Chapters 6.3.10 and 6.3.11 will introduce simpler methods which can be considered as approximations to the eddy covariance method. Deposition and emission fluxes can also be determined by the methods outlined in Chapter 6.6, while evaporation fluxes are addressed separately in Chapter 6.4.

6.3.1 Cuvettes

Biogenic emissions of trace gases from plants or parts of them may be monitored by employing cuvettes. A few leaves or a small twig are enclosed in an airtight container made from glass or plastic in order to disturb radiation fluxes as little as possible (Fig. 53). These cuvettes may be a few tens of centimetres long and a few centimetres wide. Before the start of the measurement, the container is flushed with clean air. During the measurement trace gases such as isoprene (see, e.g., Brüggemann & Schnitzler 2008) or other volatile organic compounds emitted from the plant accumulate in the container. Thus, sufficiently high trace gas concentrations for the subsequent analysis can be obtained. The emission rate is finally determined from the concentration increase divided by the length of the accumulation period. The total emission from the plant as a whole can be estimated if the enclosed leaves or twigs are considered as representative for the whole plant.

6.3.2 Surface chambers

Measurements of soil emission or absorption trace gas fluxes can be performed by a similar method to the cuvette method described above by using surface chambers (Fig. 54) made from plexiglas (Butterbach-Bahl et al. 2002). Depending on the aim of the measurements, vegetation covering the soil (grass and herbs) is either removed before the placing of the chambers or not. Before the start of the measurement, the chambers are flushed with clean or ambient air. Resetting of the chambers with ambient air can be achieved by opening the top lid of the chambers.

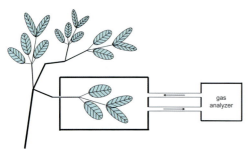

Fig. 53. Cuvette for the measurement of gas exchange from leaves and small twigs. A small pump transports the air from the cuvette to the separate gas analyzer.

6.3 Micrometeorological flux measurements

Fig. 54. Surface chambers for the measurement of soil and grass gas emissions. For the measurement, the lids will be closed. Tubes lead to gas analyzers.

Static and dynamic chambers are both currently in use (Denmead 2008). In static chambers, no air circulation takes place. These chambers do not need an external power supply. Air samples are taken at the end of the accumulation period and later analyzed in the laboratory. In dynamic chambers, there is a steady air circulation between the chamber volume and the gas analyser. They require an external power supply and the analyser must be placed close to the chambers. As long as the increasing concentration of the trace gas in the chamber during the accumulation period does not influence the emission flux, a linear increase of the trace gas concentration within the chamber can be expected. If the accumulation period is too long, then the increase of the trace gas concentration in the chamber levels off because the emission flux cannot take place against a concentration gradient. In order to avoid such saturation effects, automated dynamic chambers (Breuer et al. 2000) have been developed. These automated chambers have lids which can be opened and closed hydraulically according to a prescribed schedule. This permits a regular reset of the chambers before saturation effects can occur and a time dependent monitoring of the emission or absorption fluxes over a longer period. The chambers may be equipped with fans in order to achieve rapid mixing of the atmosphere inside the chambers as soon as the chambers are closed (Butterbach-Bahl et al. 1997). Gas analysers have been described in Chapter 5.1.

Advantages of the chamber method are its simplicity and that, due to the accumulation of the trace gases during the measurement periods, relatively simple and cheap gas analysis methods can be employed. This makes chambers suitable for longer field campaigns. Disadvantages are that the chambers change the micro-climate above the chosen surface element. Temperature, relative humidity, and turbulence inside the chambers differ from the ambient conditions. Furthermore, the air circulation should be as moderate as possible in order to avoid differences in the ambient air pressure, because such pressure differences could considerably influence the emission fluxes (Denmead 2008). Finally, it has to be noted that due to the small surface area covered by single chambers, spatial variations in the emission flux can easily lead to differences in measured fluxes between neighbouring chambers by the order of 30 to 60 % (Denmead 2008).

6.3.3 Mass balance method

The mass balance method may be viewed as an enhancement and idealisation of the cuvette and chamber methods. The method, described, e.g., in Meixner (1994, p. 307), Lenschow (1995, Ch. 5.3.4) or Denmead (2008), is based on an integration of the budget equation for the mean trace substance concentration over a hypothetical air volume. It is solved for the deposition or emission fluxes through the lower boundary of this air volume. In contrast to the micrometeorological methods to be described in the subsequent subchapters, the mass balance method delivers absolute values without the involvement of empirical relations or constants. The thermal stratification of the air is not required as an input parameter to this method. The resulting deposition or emission fluxes are determined as an areal average over the well-defined surface element as covered by the air volume.

For the closed variant of this method, concentrations, wind speeds, and wind directions must be known on all boundaries of the integration volume. By a special choice of the volume (e.g., by orienting it along the mean wind direction and by putting the upper lid at the height of an inversion) the number of necessary input data can be reduced considerably to the two lateral inflow and outflow boundaries. If the background concentration of the wanted trace substance is known with sufficient accuracy and fluxes through the upper lid are negligible it might be sufficient to make measurements at the outflow boundary only (this latter variant is also called the open method).

6.3.4 Inferential method

Chambers are not suitable to determine the trace substance exchange over larger areas or relatively high plants such as trees. Here, an analogy to Ohm's law from electricity could help. The flux of a trace substance F_c is set proportional to the voltage and to the inverse of a resistance r. If we additionally assume that the substance is completely absorbed at the surface, the concentration, C of the substance can be assumed to be zero at the surface from which we get:

$$F_c = \frac{1}{r} C \text{ or } F_c = v_d C, \tag{6.1}$$

where the inverse of the resistance, $1/r$ is written as a deposition velocity, v_d. If this deposition velocity is known, then one concentration measurement of trace substance a few metres above the surface, through which the flux is to determined, is sufficient to infer the flux from this measurement. If the concentration near the surface cannot be assumed zero, then the concentration difference $(C - C_0)$ has to be inserted into (6.1) and the method becomes similar to the gradient method as described below.

6.3.5 Gradient method

The gradient method is based on the assumption that the turbulent flux of a trace substance, F_c is proportional to the concentration gradient of the substance along the path along which the flux is sought, and to a diffusion or exchange coefficient. The latter must be known empirically or must be determined from meteorological measurements employing a similarity theory (i.e., that the diffusivity for the trace substance is similar to the measurable heat, moisture, or momentum diffusivity). Using a similarity to the momentum flux, one can write:

$$F_c = \frac{u_*\kappa z}{\varphi}\frac{\partial C}{\partial z} = \frac{u_*\kappa}{\varphi}\frac{\partial C}{\partial \ln z} = \frac{u_*\kappa(C_2 - C_1)}{\ln(z_2/z_1) - (\psi(z_2) - \psi(z_1))}. \qquad (6.2)$$

Here, φ is a stability function ψ its integral, C_1 and C_2 are the mean concentrations at heights z_1 and z_2, κ is von Kármán's constant (roughly 0.4), while u_* denotes the friction velocity which depends on the driving wind speed and the surface roughness. We note that $u_*\kappa z$ is the diffusion coefficient for momentum. The necessary stability functions for non-neutral thermal stratification of the atmosphere are known empirically (e.g., Stull 1988, Roedel 1990, or Högström 1988). The friction velocity must be determined from parallel wind profile measurements. The fluxes are assumed to be height independent in the atmospheric surface layer. Chemical reactions taking place during the transfer of the substances from or towards the surface can considerably disturb the applicability of the gradient or the inferential method. Therefore, in some cases attempts are made to apply the method to combinations of substances which react with each other (Kramm 1995). Possible combinations could be $NO-NO_2-O_3$, $NO-NO_2-O_3-HNO_3$, $HNO_3-NH_3-NH_4NO_3$ and $NO-NO_2-O_3-HNO_3-NH_3-NH_4NO_3$.

The gradient method is applicable for substances for which a vertical concentration gradient can be determined by measurements. This may become difficult for higher wind speeds when the friction velocity u_* is increased. As the flux does not increase in the same manner, the concentration gradients get smaller which may lead to problems in detecting them precisely. Further, it must be considered that the footprints (see Ch. 6.3.9.1) of the two concentration measurements between two different heights are different. Therefore, the method can only be applied for sites with horizontal homogeneity upstream for both footprints.

6.3.6 Bowen-ratio method

The Bowen-ratio method assumes that the flux of a trace substance, F_c is directly proportional to the known flux (F_T) of heat $c_p T$, the flux F_q of moisture q, or the flux of another substance. The word Bowen ratio originally refers to the ratio between the latent and the sensible heat flux. The advantage of the Bowen-ratio method is that, in contrast to the method described before, knowledge of the thermal stratification of the air is not necessary. But, as a compensation, the shortwave radiation budget of the surface R_n and the soil heat flux G are required as input parameters to the method:

$$F_c = \frac{R_n - G}{c_p \overline{\rho} \Delta \overline{T} + L \Delta \overline{q}} \Delta C \tag{6.3}$$

or:

$$F_c = F_T \frac{\Delta C}{\Delta \overline{T}} = F_q \frac{\Delta C}{\Delta \overline{q}}, \tag{6.4}$$

where the air density is ρ and the heat of evaporation is L. The concentration difference, ΔC must be determined over the same distance as ΔT or Δq. Because of the involvement of the surface energy budget, the method is sometimes called surface energy budget method. For the measurement of the vertical concentration gradient, the same requirements apply as for the gradient method (see Ch. 6.3.5).

The Bowen-ratio method can be seen as a special variant of a tracer method in which the unknown flux of one substance is set proportional to the known flux of another substance. This employs the assumption that both substances are subject to the same diffusivity.

6.3.7 Flux variance method

The flux variance method puts the substance flux, F_C proportional to the observed variance of the concentration of this substance, σ_C. This method should not be confused with the subsequently described eddy covariance method (see, e.g., Foken et al. (1995) or Lenschow (1995, Ch. 5.3.4)). The method is not able to derive the direction of the flux independently. The proportionality constant between flux and variance must be determined empirically or derived from theoretical considerations. If the proportionality constant between flux and variance is equal to the one for a second substance with a known flux, then it can be stipulated that

$$F_{C2} = F_{C1}\, \sigma_{C2}/\sigma_{C1} \tag{6.5}$$

Horizontal homogeneity surrounding the measurement site is required. The influence of chemical reactions can also not be ignored. Thus, the method is only suited for inert gases for which the direction of the flux is already known.

6.3.8 Dissipation method

The dissipation method is derived from the budget equations for turbulent kinetic energy (this equation contains a dissipation term from which this method takes its name) and for the variances of temperature, moisture, or trace substances (which likewise contain a dissipation term). The solution involves the knowledge of empirical structure functions and requires the validity of Taylor's hypothesis ($x = u\ t$). The dissipation is determined by analyzing the inertial range of measured spectra. An overview of this method and new approaches for the solution is described in Hsieh & Katul (1997).

6.3.9 Eddy covariance method

The eddy covariance (EC) method is the only direct method to determine turbulent fluxes of properties such as momentum, energy, moisture, and trace substances. The definition of the turbulent flux, the covariance of the simultaneous wind and property fluctuations, has been defined in equ. (2.2) in Chapter 2.1.5. The wind, momentum, and temperature fluctuations are usually measured with a sonic anemometer (Fig. 28), while the fluctuations of further properties must be measured with the same time resolution at a position as close by as possible without disturbing the flow conditions. Typically, the spatial distance between the centre of the wind fluctuation measurements and the property fluctuation measurement is about 30 cm. An open-path device for the detection of the property fluctuations is often mounted underneath the sonic anemometer. It could also be mounted aside the sonic but in any case it should be mounted a little bit downwind of the sonic. If the property fluctuation measurement device is a closed-path instrument, the openings of the tubes leading to this instrument should be mounted just downwind of the sonic at the same height. The method is sometimes also called the eddy correlation method, but we follow here from Lee et al. (2004), Foken (2008), Denmead (2008), and many others who all call it the eddy covariance method.

This direct method can only be applied to the fluxes of such trace substances for which high-resolution measurements in the time domain are possible (see Tab. 20).

For the measurement of CO_2 (Fig. 55) and water vapour fluxes, both open-path (e.g. LiCor 7500) and closed-path (e.g. LiCor 6262) instruments based on infrared gas analysers (IRGA) are available which offer a time resolution between 1 and 10 Hz. The open-path instrument emits an infrared light (4.26 µm for CO_2 or 2.59 µm for water vapour) beam of 1 cm diameter over a 12.5 cm long path from the main part of the instrument to the lead selenide detector mounted in the small head of the instrument. With the closed-path instrument, the air is pumped through the instrument at a rate of 6 to 10 l/min. The pumping and filters which retain dust particles lead to an attenuation of high frequencies. Therefore, correction factors must be applied (Bernhofer et al. 2003). The UV (or Krypton) hygrometer is an open path method, but with a much shorter path length of a few millimetres up to about 1 cm

Table 20. High-resolution measurement techniques for eddy covariance flux measurements.

trace substance	measurement device	reference
water vapour	infrared hygrometer, at 2.59 µm, open-path and closed-path instruments	Ch. 3.2.1.1, Moncrieff et al. (1997)
	UV hygrometer	Ch. 3.2.1.1, Fig. 56
ozone	optical gas analyser	Foken (2008)
nitric oxides	optical gas analyser, chemilumescence	Foken (2008), Rummel et al. (2002)
peroxyacetyl nitrate	thermal dissociation – chemical ionization mass spectrometer (TD-CIMS) (Wolfe et al. 2007)	Wolfe et al. (2009)
sulphur dioxide	optical gas analyser	Foken (2008)
carbon dioxide	infrared gas analyser at 4.26 µm, open-path and closed-path instruments	Moncrieff et al. (1997), Fig. 55
isoprene	chemiluminescence with reactant ozone	Guenther and Hills (1998)

(Fig. 56). The other gases listed in Tab. 20 are detected by closed-path methods, i.e. the gas is sampled very close to the sonic anemometer and is then pumped through Teflon-coated tubes to the gas analyser installed a few metres away. The development of eddy covariance methods for further substances than those listed in Tab. 20 is still ongoing. Very promising is, e.g., the use of tunable diode lasers (TDL, see also Ch. 5.1.1.1) which have already be employed for the measurements of NO_2 fluxes (Werle et al. 1993) and of methane fluxes (Werle & Kormann 2001).

Fig. 55. Eddy covariance measurement of CO_2 fluxes with an open-path gas analyser (left) and a sonic anemometer (right) (Photo: Helmut Mayer).

6.3 Micrometeorological flux measurements

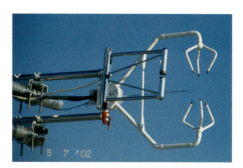

Fig. 56. Eddy covariance measurement of water vapour fluxes with a UV hygrometer (left) and a sonic anemometer (right) (Photo: Helmut Mayer).

A good representation of near-surface turbulence, which is the driving force for turbulent substance fluxes, usually requires measurements with at least a 10 Hz time resolution in order to fully capture turbulent diffusion processes. If the time resolution is worse then the high-frequency parts of the turbulent fluxes are not completely captured. On the other hand, the eddy covariance must be determined by time integration over about half an hour in order to detect the low-frequency parts correctly. Ideally, wind fluctuation and concentration fluctuation spectra and the respective co-spectrum are determined beforehand and used to fix the necessary time resolution for the flux measurements. The wind fluctuation spectrum is height-dependent. With increasing distance to the ground, turbulence elements become larger and the peak of the turbulence spectrum is shifted to lower frequencies. Thus, for larger heights above the ground, measurement rates less than 10 Hz may be sufficient for eddy covariance flux measurements. Methods to determine the limits of this method and corrections which could be applied are mentioned, e.g., in Lee et al. (2004) and Foken (2008).

One of the technical challenges of the eddy covariance method is to place wind fluctuation measurements and substance fluctuation measurements at nearly the same location without disturbing each other. When pumping gas samples through tubes into closed gas analysers, the time delay caused by the travelling time of the sample in the tube must be considered. Additionally, possible modifications of the concentration fluctuations (suppression of high-frequency fluctuations) must be accounted for.

Post-processing and quality control of eddy covariance measurements is described in detail in Mauder et al. (2006) using the TK2 software. A comparison of different post-processing software is presented in Mauder et al. (2008).

6.3.9.1 Footprint

The fluctuations measured within a few metres above ground at a micrometeorological mast are a function of the surface characteristics such as land use, roughness, and soil heat capacity upstream of the measurement site. Therefore, a flux determined by the eddy covariance method is representative for a certain area called the footprint. The size and location of the footprint could be estimated from backward diffusion modelling. If no detailed information is available, a rough first guess of the upstream extent of the footprint is one hundred times the height above ground of the eddy covariance measurement. The extent is modified by the thermal stability of the surface layer. For unstable stratification the footprint is closer the site of the measurement, while for stable stratification it is further away (further information on the determination of the footprint can be gained from Schmid 1994). To guarantee a good representativity of an eddy covariance flux measurement for a certain surface type, the footprint should be horizontally homogeneous. The transfer of the footprint concept to inhomogeneous terrain is discussed in Schmid (2002).

6.3.9.2 Webb correction

Webb et al. (1980) have demonstrated that the sensible heat flux, $\overline{w't'}$ as well as the latent heat (or moisture) flux, $\overline{w'\rho_v'}$ contribute to the measured density flux of a trace substance $\overline{w'\rho_c'}$. Because the mean vertical mass flux of air over a level surface must vanish (no mean mass flux can penetrate through the soil surface underneath) and warmer and moister rising air parcels have a lower density than descending cooler and drier air parcels, the upward motions must be slightly faster than the downward motions. Therefore, the covariance, $\overline{w'\rho_c'}$ is not exactly zero even if the substance density fluctuations associated with rising and descending motions are equal. Consequently, in order to derive the true substance flux the Webb correction must be applied to the observed heat and moisture fluxes. Following Lenschow (1995) (and for the pressure correlation term Massman 2004), this correction, which is also often termed WPL correction from the names of all three authors in Webb et al. (1980), can be written as

$$F_c \cong \overline{w'\rho_c'} + a \frac{\overline{\rho_c}}{\rho_a} \overline{w'\rho_v'} + \overline{\rho_c}\left(1 + a \frac{\overline{\rho_v}}{\rho_a}\right)\left(\frac{\overline{w'T'}}{\overline{T}} - \frac{\overline{w'p'}}{\overline{p}}\right) \qquad (6.6)$$

where a is the ratio of the molecular weights of dry air and water vapour and ρ_a the density of dry air. The correction is not required if the flux is derived via the mixing ratio, c of the trace substance with respect to dry air, i.e. $F_c = \overline{\rho} \, \overline{w'c'}$ with $c = \overline{\rho_c}/\overline{\rho_a}$. This means that for correct flux measurements without correction the air must be dried and tempered to constant temperature, which is obviously unrealistic, or that mixing ratios with respect to dry air have to be used in the eddy covariance method. If a mixing ratio with respect to moist air is used then the flux has merely to be corrected for the influence of the moisture flux. The correction for the moisture flux is only 20 % of the correction for the heat flux for equal heat and moisture fluxes.

The correction turns out to be important especially for small fluxes with $\overline{w'\rho_c'}$ < 10^{-2} $ms^{-1} \overline{\rho_c}$ or when the turbulent fluctuations of the trace substance are small against the mean concentration of this substance. The latter occurs, e.g., for CO_2 fluxes (Foken 2008). Here, and for other long-living trace gases such as CH_4 and N_2O, the correction can be of the same magnitude than the fluxes themselves (Wesely et al. 1989). Walton (1996) gives the following approximation for the WPL correction:

6.3 Micrometeorological flux measurements

$$F_c = \overline{w'\rho_c'} + \frac{\overline{\rho_c}}{\overline{\rho_a}}(0.649*10^{-6} LH + 3.358*10^{-6} SH), \quad (6.7)$$

where LH and SH denote the turbulent latent and sensible heat fluxes in W m^{-2}, respectively and the trace substance fluxes given in kg m^{-2} s^{-1}. For a critical review of the WPL correction and for further modifications which must be accounted for in open-path and in closed-path measurements, see Leuning (2004) and Massman (2004).

6.3.10 Eddy accumulation methods

For trace substances for which a high-resolution measurement of about 10 Hz is not feasible, an approximate method has been developed called the eddy accumulation method. This method is an example for a conditional sampling technique and dates back to an idea formulated by Desjardins (1977). Conditional sampling means that the measurement procedure is governed by a further variable. In the eddy accumulation method, a trace substance is collected into two different containers depending on the sign of the vertical wind fluctuation (denoted by w$^+$ und w$^-$). The intake rate at the two containers depends on the absolute value of the vertical wind speed fluctuation. The masses, c_1 and c_2 of the trace substance in the two containers are analyzed after a pre-defined time interval. The calculated mass difference is proportional to the unknown flux:

$$F_c = \overline{w^+c^+} - \overline{w^-c^-} \propto c_1 - c_2. \quad (6.8)$$

The technical challenge in the application of this method is the accurate triggering of the valve which controls the flow in either of the two containers depending on the sign and size of the vertical wind fluctuations.

The employment of the pure eddy accumulation method is additionally problematic because usually the mass difference in the two containers is quite small, and because it is difficult to remove any systematic errors from the measurement of the mean vertical velocity. Therefore, the relaxed eddy accumulation method (see below) has been proposed by Businger & Oncley (1990).

6.3.10.1 Relaxed eddy accumulation method

The relaxed eddy accumulation (REA) method involves the evaluation of the following relation:

$$\overline{w'c'} = b\sigma_w \left(\overline{c^+}\ (w > w_0) - \overline{c^-}\ (w < w_0)\right), \quad (6.9)$$

which in contrast to (6.8) contains the empirical constant b (~0,6). In contrast to the pure eddy accumulation method, the air is directed into one of the two containers vice. The Piche evaporimeter has, as an evaporating element, a disk of filter paper attached to the underside of an inverted and graduated cylindrical tube, which is closed at one end and supplies water to the disk at the other. Successive measurements of the volume of water remaining in the graduated tube will give the amount lost by evaporation in any given time (WMO 2006).

6.4.2 Lysimeters

Evapotranspiration from grassland may be continuously monitored by a lysimeter. A lysimeter is a cylindrical tank lowered into the earth which is filled with the undisturbed soil core extracted from the ground to prepare the hole for the lysimeter. They are installed so that the grass surface in the lysimeter is at the same level as the surrounding grass surface. According to their method of operation, lysimeters can be classified into non-weighable and weighable instruments. Non-weighable lysimeters are for long-term observations only. The percolated water is measured and compared to the precipitation amount fallen on the lysimeter. Weighable lysimeters are placed on a weighing mechanism (Pruitt & Angus 1960). The slit between the tank and the surrounding soil should only be a few millimetres wide (see Figs. 57 and 58). There are no standard lysimeters (WMO 2006). A typical lysimeter tank is about one to two metres deep (depths range from 0.1 to 5 m) and about half a metre wide (widths range from 0.1 to 10 m). Excess leachate water can be let off from the bottom of the tank in order to keep soil moisture con-

Fig. 57. Lysimeter for evaporation measurements (see Fig. 58 for a schematic). A rain gauge at equal level with the surrounding surface is seen in the background. Splashing of water from the surrounding surface into the gauge is reduced by the grid around this rain gauge (Photo: Helmut Mayer).

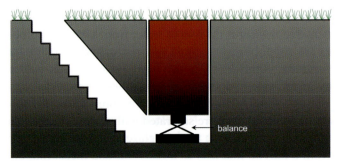

Fig. 58. Schematic of a weighable Lysimeter (see also Fig. 57).

ditions comparable to the conditions in the surroundings. The weight changes of the lysimeter are proportional to precipitation and evaporation. The actual evaporation can be measured, if the precipitation amount is determined in parallel by a rain gauge (Ch. 4.1). The potential evaporation can be measured, if the lysimeter is watered to keep the soil moisture steadily at the soil field capacity (see e.g. Parlange & Katul 1992). The accuracy of a lysimeter is about 0.03 mm of equivalent water depth (Pruitt & Angus 1960).

6.4.3 Evaporation pans and tanks

Potential evaporation can most easily be measured by monitoring the water level in standardized water tanks. The most well known tank is the class A pan, a round water container from stainless steel with a diameter of about 1.2 m (surface area about 1.14 m^2) and a water depth of 0.2 m (see Fig. 59). The adoption of the Russian 20 m^2 tank (about 5 m in diameter and 2 m deep) as the international reference evaporimeter has been recommended by WMO (2006). If pans are exposed for longer times, precipitation has to be measured separately. An automated evaporation pan is described in Chow (1994). The evaporation from larger water surfaces can be estimated with pans mounted on a small raft. An overview of evaporation measurements from open water surfaces is given in Vietinghoff (2000).

6.4.4 Recommendations for evaporation measurements

Prior to the development of micrometeorological methods atmometers and evaporimeters were the only means to directly determine the moisture flux from the surface into the atmosphere. In contrast to the footprint-dominated micrometeorological methods described in Chapter 6.3, point measurements are performed with atmometers and evaporimeters. The transferability of the results to larger areas has to be checked carefully.

Atmometers are sensitive to dirt on their evaporating surfaces and must be kept clean. Pans sunken in the ground seem to be most ideal due to energetic reasons (e.g., they are not cooled at their lower side by the wind), but can be subject to splashing from the surroundings and to pollution with leaves and other unwanted materials. Pans mounted some centimetres above ground are less prone to splashing and pollution but their results can be modified by radiant energy intercepted by the sides. Lysimeters must be kept in the same state as their surroundings which can be difficult due to the restricted growth of the plant roots in the container, thermal insulation from the surround-

Fig. 59. Evaporation pan for evaporation measurements (Photo: Helmut Mayer).

ing soil and restricted drainage. Pans and lysimeters have to be checked regularly for leaks.

Technical rules for the determination of evaporation are published in Appendix A of the VDI guideline 3786 part 13 (see Appendix). Detailed description of evaporation measurements can be found in Chapter 7 of Strangeways (2003) and Chapter 10 of WMO (2006).

6.5 Soil heat flux

The soil heat flux can contribute up to 10 % of the instantaneous surface energy balance but vanishes on a yearly average. The soil heat flux can be recorded by a heat-flux plate, a plate which is about 10 cm in diameter and a few millimetres thick. This plate has a comparable heat conductivity to that of the soil and is burrowed in the soil. The temperature difference between the upper and the lower side of the plate is measured with thermistors (Ch. 3.1.3) or thermocouples (Ch. 3.1.4) and the soil heat flux is computed from this difference via a flux-gradient relationship. The theoretical basis of this method is described in Philip (1961), a self-calibrating flux plate is presented in Liebethal & Foken (2006) in more detail. Problems can arise when disturbing the soil structure while burrowing the plates, from mismatch between the thermal conductivity of the plate to that of the soil, and from insufficient thermal contact between the plate and the soil (Strangeways 2003).

6.6 Inverse emission flux modelling

When emission fluxes from sources are not well-defined (diffuse sources) or come from inaccessible sources, placement of a single or only a few instruments will not be effective. In that case, fluxes can be determined from inverse modelling techniques. These techniques consist of downwind concentration measurements of the emitted trace substance and a backward numerical simulation of the dispersion. For ubiquitous substances, the background concentration upwind of the source must be determined, too. In some respect the relationship between the emission source strength and the downwind concentration is comparable but inverse to the relationship between the upwind footprint and the micrometeorological flux measurements. In order to cope with inhomogeneities in the dispersed plume downwind of the source, path-averaged concentration measurements (see Ch. 7.3.2 and 7.3.3 below) are advisable. The principal outline of this method is sketched in Fig. 60. A special inverse method is the Backward-Lagrange method described in Flesch et al. (1995). Two monographs on inverse modelling that are not restricted to isolated pollution sources but also deal with global methane and carbon dioxide emissions, are the

6.6 Inverse emission flux modelling

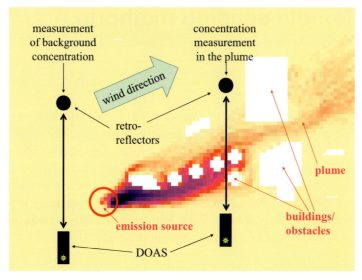

Fig. 60. Principal outline of an inverse emission source strength determination from the difference between an upstream and a downstream path-averaged concentration measurement with a DOAS (see Fig. 90).

books by Bennett (2002) and Enting (2002). The success of this method decisively depends on the quality of the turbulence parameterization used in the numerical dispersion models (Denmead 2008).

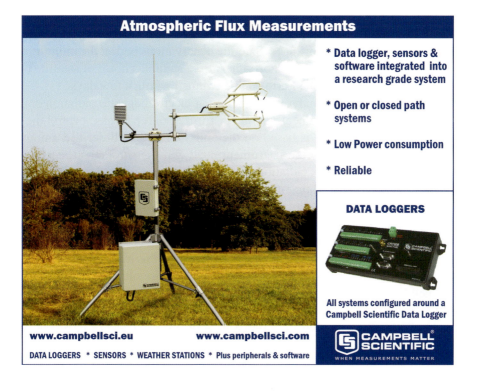

7 Remote sensing methods

Chapter 7 is structured differently to Chapters 3 to 6. We will start here with a description of the different available remote sensing techniques. The subsequent Chapters 8 to 12 will then be structured according to the measured variables as before. This is advisable since all remote sensing methods are based on a few fundamental principles such as the absorption, emission, scatter, or reflection of electromagnetic or acoustic radiation. Selected variables are then derived from specific properties of radiation transfer in the atmosphere. Some variables can only be measured with the combined use of more than one remote sensing principle.

7.1 Basics of remote sensing

Surface-based measurements and in-situ observations from platforms such as masts or aircraft (see Ch. 2.5) only give very limited data coverage for the free atmosphere. These disadvantages have led to the development of remote sensing methods operated either from the ground, from aircraft or from satellites. The progress in satellite technology has especially driven this development because remote sensing is the only way to receive information on the state of the atmosphere and the underlying ground or water surfaces from space. Satellite observations permit the monitoring of large parts of the Earth surface in short times. By a suitable choice of satellite orbits nearly the whole surface of the Earth can be scanned within a few days.

Passive remote sensing analyses the incoming electromagnetic radiation emitted from either within a measurement volume or from a source located behind the volume. Active remote sensing is based on the emission of a well-defined amount of electromagnetic radiation by the instrument itself and the subsequent reception of the backscattered part of this radiation. If the receiver is combined with the emitter, the method is called monostatic, otherwise bistatic (see Fig. 61).

Remote sensing results are influenced by the properties of the emitting source as well as by the medium through which the radiation must pass. A separation between both contributions is only possible by complex inversion algorithms which usually involve the radiation transfer equation (RTE). Formally, this inversion procedure can be described as follows: an object O_1 (for example, an air volume) emits a signal S_{a1} determined by a collection of properties we denote here as E_1. The relation between S_{a1} and E_1 is described by a (not necessarily linear) function g: $S_{a1} = g(E_1)$. Emitted thermal radiation from an object is a function of the fourth power of its temperature. The receiver sees a signal, S_e coming from this object. This signal, S_e is usually different from the signal, S_{a1} that has been emitted from O_1 because it has been modi-

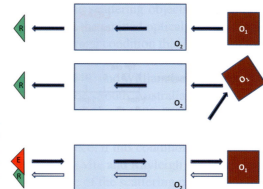

Fig. 61. Principal geometries of remote sensing. Top: passive remote sensing of the properties of object O_2 using the radiation emitted from an object O_1. Middle: as above, but O_1 is just reflecting the radiation from another source. Below: active remote sensing of O_2 by using the reflection at O_1.

fied when passing through an object O_2 with a collection of properties denoted as E_2 (usually a part of the atmosphere) in between. Thus

$$S_e = f(S_{al} = g(E_1), E_2), \qquad (7.1)$$

where f is a nonlinear transfer function (e.g., the RTE) of the received signal. The main task of inversion algorithms for remote sensing data is either:

a) to deduce the radiation S_{al} emitted or reflected from O_1 from S_e and thus determine E_1 from

$$E_1 = g^{-1}(S_{al}) = g^{-1}(f^{-1}(S_e, E_2)) \qquad (7.2)$$

or

b) to deduce E_2 from the received signal S_e according to

$$E_2 = f^{-1}(S_e, S_{al} = g^{-1}(E_1)). \qquad (7.3)$$

The task is thus to find the correct inverse function f^{-1}. The function g and its inverse are usually assumed to be known. The main problem is that this inversion task in most cases does not have a unique solution; i.e., different states of the objects O_1 and O_2 can lead to the same received signal S_e (Liou 2002). The only way out from this problem is the assumption of a first guess (Bayesian method) or at least some additional constraints to the solution. The solution of (7.3) is relatively simple if either (a) the radiation is not modified between the object and the receiver ($S_{al} = S_e$) or (b) the emitted radition S_{al} is known. An example for the situation (a) is observation from space where the radiation propagation in the vacuum of space is undisturbed. Situation (b), e.g., is generated when the sun or another known radiation source is used in absorption spectroscopy.

Table 22. Basic remote sensing techniques.

name	principle	spatial resolution	direction	type
RADAR	backscatter, electro-magnetic pulses, fixed wave length	profiling	scanning, slanted	active, monostatic
SODAR	backscatter, acoustic pulses, fixed wave length	profiling	fixed, slanted, vertical	active, usually monostatic
LIDAR	backscatter, optical pulses, fixed wave length(s)	profiling	scanning, fixed, horizontal, slanted, vertical	active, monostatic
RASS	backscatter, acoustic, electro-magnetic, fixed wave length	profiling	fixed, vertical	active, monostatic
FTIR	absorption, infrared, spectrum	path-averaging	fixed, horizontal, slanted	active, bistatic or passive
	emission, infrared, spectrum	path-averaging	fixed, horizontal, slanted	passive
DOAS	absorption, optical, fixed wave lengths	path-averaging	fixed, horizontal	active, bistatic
radiometry	electro-magnetic, fixed wave length(s)	averaging, profiling	fixed, scanning, slanted, vertical	passive
tomo-graphy	travel time, acoustic, fixed wave length	horizontal distribution	fixed, horizontal	active, multiple emitters and receivers

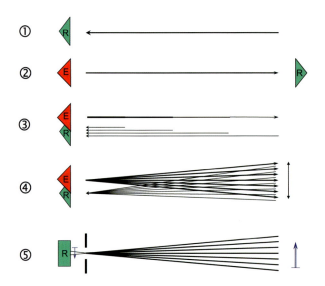

Fig. 63. Principal types of remote sensing. (1): path-averaging passive remote sensing, (2): path-averaging active bistatic remote sensing, (3): sounding, (4): scanning, (5): imaging. "R" denotes receiver, "E" emitter.

cal photography, delivers spatial distributions of heat sources, clouds, surface properties, etc. Sounding delivers a distribution of a variable along the line of sight, whereas path-averaging yields just one average value for the whole line of sight. Scanning refers to sounding or path-averaging at different angles in a given sequence.

7.2 Active sounding methods

Active sounding remote sensing methods analyse the backscattered parts of pulses with known wavelength and intensity emitted by the instruments into the atmosphere. The distance of the backscattering atmospheric volume is determined in most applications of this measurement method from the travel time of the signal. In addition to the backscatter intensity, many active remote sensing techniques also analyse the Doppler shift of the backscattered signal to determine the velocity of the backscattering object parallel to the line of sight (radial velocity component).

The Doppler shift Δf (named after the Austrian physicist C. Doppler (1803–1853)) is proportional to the emitted frequency f_0 and to the radial speed of the scattering object, v when

$$\Delta f \approx 2 f_0 v / c. \tag{7.4}$$

The relative frequency shift, $\Delta f/f_0$ depends only on the ratio between the velocity of the scattering object and the propagation speed of the emitted signal. The received wavelength is reduced if the air volume moves towards the instrument and it is increased if the air moves away.

7.2.1 RADAR

RADAR (radio detection and ranging) is the oldest active remote sensing technique (Fig. 64). It analyses the Rayleigh backscatter of electro-magnetic waves from hydrometeors (Wexler & Swingle 1946). But occasionally swarms of insects and birds lead to considerable echoes, too. Bragg backscatter from clear air is not usable unless strong moisture gradients – mainly in clouds – are present (Knight & Miller 1998). Therefore, the RADAR technique has been developed to measure the location and intensity of precipitation (weather RADAR). It can also be used for horizontal wind measurements, if the Doppler shift of the backscattered radiation is analyzed (Doppler-RADAR).

Fig. 64. Radom of the weather RADAR of the German Weather Service at the Hohenpeißenberg observatory.

Table 23. RADAR wave lengths and frequency bands.

frequency	wave length	frequency bands	application
20–300 MHz	1–15 m	VHF	windprofiler
400–900 MHz	30–70 cm	P-Band/UHF	windprofiler
1–2 GHz	15–30 cm	L-Band/UHF	boundary layer windprofiler
2–4 GHz	7–15 cm	S-Band/UHF	precipitation
4–8 GHz	4–7 cm	C-Band	precipitation
8–16 GHz	2–4 cm	X-Band	precipitation
16–20 GHz	1–2 cm	Ku-Band	precipitation
35 GHz	8,5 mm	Ka-Band	precipitation, clouds
90–100 GHz	3 mm	W-Band	clouds

Typically eight wave length bands (see Tab. 23) are differentiated in RADAR meteorology. L and P band and VHF RADAR devices are called windprofilers (see next subchapter). The K band is sometimes divided into Ka and Ku bands. A guide to interpretation of RADAR images is found in Bader et al. (1995). Many aspects of the application of RADAR in meteorology are collected in Atlas (1990) and Wakimoto & Srivastava (2003).

A RADAR consists of a parabolic antenna with an emitter and receiver mounted in its focal point. The antenna serves both to emit and receive the signal. Because the aperture of the antenna and the RADAR wave length are nearly of equal size, beam refraction is important and the RADAR beam cannot be focussed perfectly. During the emission, side lobes are produced, i.e., a non-negligible part of the radiated energy is emitted at larger angles away from the main beam. Backscattered energy from these side lobes leads to backscatter from the ground (ground clutter) and from obstacles away from the focal line (fixed echoes) which can disturb the measurements considerably. Known fixed echoes can be eliminated a posteriori. Strong electro-magnetic radiation from a RADAR is dangerous to animals and human beings because it heats up body cells and can damage them. Therefore, one has to keep away from the direct vicinity of the antenna.

The ratio between emitted and received power is described by the RADAR equation:

$$P_R = r^2 \frac{c\tau A \varepsilon}{2} P_0 \kappa^2 Z e^{-2\sigma r} + P_{bg}. \tag{7.5}$$

Here, P_0 denotes the emitted power (usually some hundreds of kW), P_R is the received power (minimum detectable power is about 10^{-14} W), P_{bg} is the power of the background noise, ε is the efficiency of the antenna, A is the effective area of the antenna, σ is the scattering and absorption by particles soaring in the air, r is the distance to the RADAR, τ is the pulse duration (typically half a micro second), $\kappa^2 Z$ is the backscatter cross-section ($\kappa^2_{water} = 0.93$, $\kappa^2_{ice} = 0.17$), and c is the speed of

7.2 Active sounding methods

light. The ratio of the two terms on the right-hand side of the RADAR equation is called signal-to-noise ratio (SNR). For the derivation of this equation, Rayleigh scattering has been assumed. Therefore, the wavelength of the RADAR, λ must be much larger than the diameter, d of the precipitation particles ($d < 0.05\ \lambda$ for water droplets and $d < 0.1\ \lambda$ for snow flakes). This equation gives an integrated value over all precipitation particles in the scattering volume. The range resolution Δr is given by:

$$\Delta r = 0.5 c \tau. \tag{7.6}$$

(7.6) gives a range resolution of 75 m when the pulse duration is half a micro second. Because the received backscattered power is also proportional to the pulse duration, a trade off between maximum range and range resolution has to be made.

Usually RADAR beams are emitted nearly horizontally or at a small angle to the horizontal plane. Therefore, the refraction of the beams is very important when computing their path length. Refraction can only be neglected for vertical beams. The magnitude of the refraction depends on the vertical gradient of the refraction index, n of the atmosphere for electro-magnetic radiation. This refraction index is mainly a function of the temperature and moisture distribution. In order to get convenient expressions, often a refractivity N is defined by

$$N = 10^6 (n - 1). \tag{7.7}$$

The temperature and moisture dependence of this refractivity can be expressed as:

$$N = 77.6 \frac{p}{T} + 3.73 \cdot 10^5 \frac{e}{T^2}, \tag{7.8}$$

with the air temperature, T in K, the air pressure is p and the water vapour pressure is e in hPa (Bean & Dutton 1968). An interesting problem concerning the refraction of the RADAR beam near the surface is whether the curvature of the beam is smaller or larger than the curvature of the Earth's surface. For this purpose, a modified refractivity M is defined by

$$M(z) = 10^6 \frac{z}{R} + N = 157\ z + N, \tag{7.9}$$

where the height above the Earth's surface is z and the Earth's radius is R, both given in km. If $M(z)$ turns out to be constant with height, z then the RADAR beam follows the curvature of the Earth's surface always having a constant height above ground. If $M(z)$ decreases with height, the RADAR beam is bended towards the ground. In layers with a negative vertical gradient of M, RADAR beams are captured like in a wave duct. Such ducts can occur underneath of strong inversions. Vertical gradients of M between 0 and 78 km^{-1} are called superrefraction. Such conditions appear with stable thermal stratification and a strong vertical decrease of moisture, and they lead to enhanced maximum ranges of a RADAR because the height above ground of the RADAR beam is only increasing slowly. Gradients of M between 78 and 157 km^{-1}

are defined as normal propagation conditions and gradients above 157 km^{-1} as subrefraction. Under subrefraction conditions, the RADAR beam is bended upward more than normally and the maximum range of the instrument is reduced because the beam is too high above ground in larger distances from the instrument. A climatology of refraction conditions for the Alps is given in Viher (2006).

7.2.1.1 Micro rain RADAR

Also based on the RADAR technology are micro rain RADARs (Fig. 65). They are frequency-modulated continuous-wave (FM-CW) RADAR instruments which emit a vertically directed beam with a wavelength of 1.25 cm (24.1 GHz). Micro rain RADARs record not only the backscattered intensity (reflectivity) but also analyse via the Doppler shift of the backscattered signal the falling velocity of the scattering precipitation particles. With this coinciding information on fall speed and drop size, a quantitative precipitation measurement is possible (Löffler-Mang et al. 1999, Peters et al. 2002, 2005). The instrument has an emitting power of 50 mW and can resolve 28 height range intervals between 35 and 200 m. If a height interval of 100 m is chosen, a layer from 300 m to 3000 m above ground can be analyzed. The minimum time resolution is 10 s; typically 60 s is used. The small instrument which weighs just 12 kg operates with a parabolic antenna of 60 cm diameter that emits the 2° wide beam. The device is usually mounted on a mast of a few metres height and does not require shielding by a radom.

7.2.1.2 Synthetic-Apertur RADAR

A Synthetic-Aperture RADAR (SAR) employs an extensive procedure, which is comparable to holography, to deliver photographic images in the VHF range (Tab. 23) that are not disturbed by the presence of clouds. In this interferometric procedure, amplitude and phase information from several subsequent satellite RADAR images are brought together. The artificially large aperture of the sensor is due to the movement of the satellite during the reception of data. This method has also been used to analyse the surface of the completely clouded planet venus. SAR is used for

Fig. 65. Micro rain RADAR operating at 24.1 GHz. (Photo: METEK GmbH)

7.2 Active sounding methods

Earth observations on the satellite ERS-1 since 1991. Issues, techniques and applications of SAR interferometry are reviewed in Gens & van Genderen (1996).

7.2.1.3 Altimeters

Altimeters (Latin: height meters) are used for the determination of the altitude of the Earth surface, especially sea surfaces (Wunsch & Stammer 1998). They are based on the precise measurement of the distance between the sensor (mounted on a satellite or on an aircraft) and the Earth surface. Altimeters emit short pulses in the VHF range and record the travel time from the time interval passing between the emission of the pulse and the reception of the backscattered signal. A necessary prerequisite for the operation of an altimeter is the continuous knowledge of the exact position of the satellite or aircraft. Altimetry may also be employed to detect and quantify gravity anomalies of the Earth (Sandwell & Smith 1997).

7.2.1.4 Scatterometers

Scatterometers on satellites measure the intensity of backscattered VHF-RADAR pulses at varying angles. This intensity is strongly determined by the roughness and structure of the backscattering surface. Because it is difficult to really focus RADAR beams the spatial resolution of scatterometers is rather crude (about 50 km). Recent progress in the scientific application of space-based scatterometer data is reviewed by Liu (2002).

7.2.2 Windprofilers

Windprofilers are Doppler-RADAR instruments for the detection of the vertical wind profile working at frequencies of 50 MHz to 1 GHz (VDI 3786 part 17, see Appendix). The respective wave lengths are between 6 m and 0.3 m. Windprofilers operating at up to about 300 MHz are called VHF windprofiler (very high frequency). Those operating at higher frequencies are UHF windprofilers (ultra high frequency) (see Tab. 23). In this frequency range, we observe usable Bragg backscatter of temperature and especially of moisture fluctuations in the atmosphere. Water and rain drops do not absorb in this range, therefore Windprofiler measurements are not disturbed by the presence of clouds and rain. VHF windprofilers give profiles from the stratosphere and mesosphere (Fig. 68), UHF windprofilers have a lower range and are used for the observation of the troposphere (Figs. 66 and 67). A recent overview of this technique is given in Muschinski et al. (2005).

Due to the high propagation speed of electro-magnetic waves (in the order of $3 \cdot 10^8$ m/s), the height resolution of windprofilers is only about 100 m. The range depends on the chosen frequency (from about 1 km at 1 GHz to about 70 km at 50 MHz) and the emitted power. The size of the antenna is proportional to the wave length. For 6 m wavelengths an area of about 100 m times 100 m is required, for

Fig. 66. Boundary-layer wind profiler (left) with acoustic RASS extension (right) of the German Weather Service at the Lindenberg observatory.

Fig. 67. Tropospheric wind profiler with radiation shield around it and acoustic RASS extensions (yellow cylindrical objects to the inside of the shield) of the German Weather Service at the Lindenberg observatory.

Fig. 68. Antennas of the VHF wind profiler OSWIN that delivers wind profiles from the stratosphere and mesosphere of the Leibniz Institute for Atmospheric Physics at Kühlungsborn.

shorter wavelengths 10 m by 10 m with yagi elements or even a bowl-shaped antenna may be sufficient (Strangeways 2003).

The first windprofilers were a 40 MHz profiler in Boulder, Colorado in 1975 and a 53.5 MHz profiler near Katlenburg-Lindau, Germany in 1976. Today, there are more than 100 windprofilers in the United States. The German Weather Service operates a 10 m by 10 m large 482 MHz windprofiler with a range of 16 km at its observatory in Lindenberg (Fig. 67). The pulse duration is typically from a few microseconds to a few tens of microseconds; the emitted power is between 16 and 18 kW. This windprofiler has been upgraded to a RASS (see Ch. 7.2.4 below) by an addition

7.2 Active sounding methods

of an acoustic emitter and is thus able to measure temperature profiles up to a height of about 4 km. The acoustic frequency has been chosen to be half of the electro-magnetic wave-length, i.e. the Bragg condition is obeyed. The instrument is shielded by a high metal fence against interaction with nearby television emitters. At the same observatory, a 1290 MHz boundary layer windprofiler operates (Fig. 66). It has a range of 4 to 6 km and has likewise been extended to a RASS by the addition of an acoustic emitter.

Like RADAR instruments, the emitted beams from windprofiles have side lobes which can lead to the recording of fixed echoes. Wind speed measurements from these instruments are disturbed by moving obstacles (swaying trees, wind energy converters). Shielding to the sides can reduce the disturbance of fixed echoes.

7.2.3 SODAR

Acoustic remote sensing devices (Fig. 69) for vertical sounding emit and receive sound pulses, and are called – in analogy to RADAR – SODAR (sound detection and ranging, Fig. 70) (Gilman et al. 1946). An overview of the basic principle (see

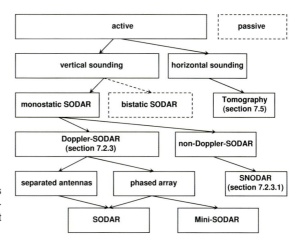

Fig. 69. Schematic overview of methods for acoustic remote sensing of the atmosphere. Items in dashed boxes are not described in this book.

Fig. 70. Movable mono-static Doppler SODAR with three antennas for the measurement of wind and turbulence profiles.

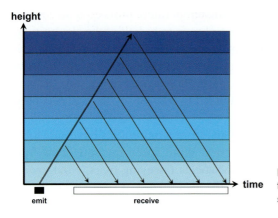

Fig. 71. Schematic of the distance determination with active remote sensing instruments such as RADAR, SODAR, and ceilometer.

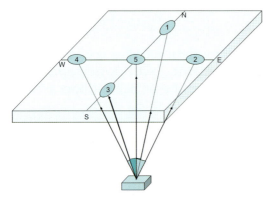

Fig. 72. Schematic of the Doppler-beam swinging (DBS) method to determine the three-dimensional wind vector. Some SODAR instruments use three beams only (beams 1, 2, and 5), other use all five beams during one measurement cycle.

sketches in Fig. 71 and 72) of sounding with a SODAR and the history of the development of this instrument is described in Peters (1991); a most recent overview on acoustic remote sensing techniques and its theoretical backgrounds is given by Bradley (2007). SODARs analysing the Doppler shift of the backscattered sound pulses for the derivation of the vertical wind profile are called Doppler-SODARs.

Sound waves are scattered by turbulent temperature fluctuations in the atmosphere that are supposed to move with the mean wind since the refraction index for sound waves is changing at the boundaries of these fluctuations. For a monostatic SODAR (the position of the receiver is identical with the one of the emitter), the intensity of the backscattered signal depends only on these temperature fluctuations, while for a bistatic SODAR (the receiver is deployed away from the emitter) velocity fluctuations in the atmosphere also contribute to the intensity of the backscattered signal. For optimal backscatter, the spatial size of the temperature gradients in the atmosphere should be about half the acoustic wavelength (Bragg condition). This condition can be fulfilled by turbulent fluctuations as well as by temperature inversions. When deducing information on the atmospheric state from the backscattered signal, one has to differentiate between these two possibilities. This can be done either from an assessment of the general weather conditions or (if the SODAR is a

Doppler-SODAR) from an analysis of the variance of the simultaneously recorded vertical velocity. High variances indicate thermal forcing and the backscatter intensity is proportional to the turbulence intensity, low variances indicate stable layering and the backscatter intensity should be proportional to mean vertical temperature gradients (i.e. inversions).

The emitted power of a SODAR is of the order of 1 kW, the backscattered power is only of the order of 10^{-15} W. The low power is less than the detection threshold of a human ear (10^{-12} W) but still several orders of magnitude above the thermal noise of the Brownian motion of the air molecules ($6·10^{-19}$ W). The ratio between emitted and received power is described by the SODAR equation:

$$P_R = r^{-2}\,(c_s \tau A \varepsilon / 2)\, P_0\, \beta_s\, e^{-2\sigma r} + P_{bg}. \tag{7.10}$$

where the received power is P_R, the emitted power is P_0, the antenna efficiency is ε, the effective antenna area is A, the sound absorption in air is σ due to classical and molecular absorption because of the collision of water molecules with the oxygen and nitrogen molecules in air, the distance between the scattering volume and the instrument is r, the pulse duration is τ (typically between 20 and 100 ms), the backscattering cross-section is β_s (typically in the order of 10^{-11} m^{-1} sr^{-1}), the sound speed is c_s, and the background noise is P_{bg}. The background noise also comprises contributions from ambient noise having the same sound frequency, e.g., traffic noise. The ratio of the two terms on the right-hand side of the SODAR equation (7.10) is called signal-to-noise ratio (usually abbreviated as SNR). The backscattering cross-section, β_s is a function of the temperature structure function, C_T^2 (Tatarskii 1961). For a monostatic SODAR, we find (Reitebuch 1999) when using the wave number $k = 2\pi/\lambda$ that

$$\beta_s(180°) = 0.00408\, k^{1/3}\, C_T^2\, /T^2. \tag{7.11}$$

This equation forms an average over all scattering elements within the atmospheric volume that is hit by the cone-shaped beam. The pulse duration, τ determines the height resolution of the instrument via the relation:

$$\Delta z = 0.5\, c\, \tau. \tag{7.12}$$

The SODAR equation (7.10) shows that the backscattered power is proportional to the pulse duration, too. Therefore, the choice of the pulse duration is a trade-off between height resolution (preferably short pulse durations) and maximum range (preferably long durations).

The backscattered signal of a SODAR is, like with a RADAR, representative of a certain atmospheric volume. Given a pulse duration of 100 ms, we have simultaneous backscatter from a volume of 33 m depth. For a height of 500 m above ground, this volume has a radius of 44 m (assuming an opening angle of the emitted sound beam of 5°). Additionally, the three-dimensional wind information has to be inte-

Fig. 73. Array of 64 sound transducers of a phased-array SODAR without baffles (i.e. sound absorbing shields). (Photo: Helmut Mayer)

Fig. 74. Phased-array SODAR. The antenna is to the inside of the baffles.

grated from a full measurement cycle over one vertical and two to four tilted beams (e.g. from the beams 1, 2, and 5 in Fig. 72). This leads to a radius of the detected atmospheric volume in 500 m height of about 150 to 200 m. This makes SODAR a volume and time averaging measurement device.

Depending on the technology how the sound beams are focussed, a difference is made between "classical" SODARs, which had been developed first, and phased-array SODARs. Classical SODARs use three large tiltable horn antennas (Fig. 70) to focus the beams. Several large simultaneously operating sound transducers are fixed at the bottom of each of these horns. The sound transducers serve as emitters as well as receivers. Phased-array SODARs (Figs. 73 and 74) have a larger number of smaller sound transducers (often 32 or 64) which are regularily arranged on a quadratic plate with an area of about 1 m^2. By using the interference principle (enhancement and extinction), slanted and vertical beams are formed by operating the various sound transducers with specified time delays among each other (Fig. 75). While with classical SODARs the antennas simultaneously serve as a shield to protect the environment from the sound pulses as well as the SODAR from disturbing ambient noise,

7.2 Active sounding methods

Fig. 75. Schematic of beam forming with a phased-array SODAR. Left: Loudspeaker #6 operates first, middle: all loudspeaker operate simultaneously, right: loudspeaker #1 operates first.

phased-array SODARs need a shield around them (Fig. 74). Nevertheless, even if optimal lateral shields are erected, the sound pulse is still detectable for human ears at a distance of several hundreds of metres. Thus, siting of SODARs close to residence areas and office buildings has to be avoided or needs the explicit approval of the people living and working there.

Doppler-SODAR instruments for the measurement of wind profiles work with the Doppler-beam-swinging (DBS) technology. Here, three to five beams are emitted one after the other into three to five different directions (Fig. 72). One of the three or five directions is always the vertical direction, the other directions are tilted by 15° to 20° from the vertical. The azimuthal directions of the tilted beams usually differ by 90°. Because the next pulse can not be emitted before the backscatter of the previously emitted beam has been recieved in order to avoid disturbances, one measurement cycle of a SODAR lasts six to ten times the time a sound pulse needs to travel up to the maximum vertical range of the instrument. For a SODAR operating at 1500 Hz, a range of 1000 m rougly means a length of a measurement cycle of about 20 s. The minimum range gate is around 20 to 30 m above the instrument; the vertical resolution depends on the pulse duration and is in the order of 5 to 20 m.

The maximum vertical range depends on the emitted power and sound frequency with ranges from about 200 m at 4500 Hz to about 1 km at 1500 Hz. The frequency dependence is due to the different absorption of sound waves of different frequency in the atmosphere. Absorption increases with frequency. 4500 Hz-SODARs with a maximum range of up to 200 m are frequently called MiniSODARs. Some experiments with high-frequency SODARs (6 to 20 kHz) for turbulence detection have been performed by Coulter & Martin (1986) and Weill et al. (1986), but have not found general application.

Usually a time average of 15 to 30 min over several cycles is calculated in order to deal with the statistical uncertainty of the results from single measurement cycles. The variance of the wind components can be determined by either the variance of the single wind measurements with one averaging period between 15 and 30 min, or better, from the width of the spectral peak in the backscattered signal.

In addition to mono-frequency SODARs, multi-frequency SODARs are also available which emit several subsequent pulses with different frequencies within one shot. Field tests have proved significant advantages of the multi-frequency technique. The use of, e.g., eight different frequencies halves the minimal acceptable signal to noise ratio compared to single-frequency sounding. Moreover, the multi-frequency mode improves the accuracy of instantaneous values of measured parameters and significantly increases the reliability in recognizing noisy echo-signals. Further details on the principle of multi-frequency SODARs are given in Kouznetsov (2009).

Further principles of pulse code methods for the enhancement of data availability and range are reviewed and investigated in Bradley (1999). In particular, detailed simulations are performed, using weather-like targets, of a comb of frequencies, a chirp, and a phase-encoding method. Three Doppler-adaptive matched filters are described, and two of these evaluated against the simulated noisy atmosphere. It is found that the comb of frequencies produces the least variance in estimated Doppler wind speed. A filter based on a single evaluation of an FFT for the received signal provides Doppler winds to about 1 %. The Doppler-adaptive filters add little computational or hardware overhead, and produce as a simple output a best estimate of the wind speed component.

SODAR measurements depend on the state of the atmosphere. If the atmosphere is extremely well mixed, i.e. temperature fluctuations are very small, nearly no sound is reflected from the atmosphere and the signal to noise ratio for SODAR can be that large that the determination of a wind speed (via the Doppler shift) is not possible (this happens most pronounced in the afternoon of days with strong vertical mixing due to thermal heating, usually days with small mean wind speeds). Further, SODAR measurements are disturbed by sound sources in the near vicinity of the instrument (this includes wind noise which is excited at the instrument itself). The latter problem limits the measurement of extremely high wind speeds.

Like RADAR beams, SODAR beams also have side lobes (see Simmons et al. 1971 for side lobe diagrams for different sound frequencies). Therefore, fixed echoes can severely influence and disturb SODAR measurements. This requires a good site selection for these instruments. SODARs have to be sited away from obstacles like buildings, trees, and electric cables. Furthermore, sites have to be several hundreds of metres away from human settlements because the regular beeps from the instrument disturb the residents. Exposure to the sound pulses directly above the antennae is dangerous to human health and has to be avoided in any case. Ear protectors must be worn when working at the antennae of a running SODAR.

7.2.3.1 SNODAR

A special SODAR development for extreme cold environments is a SNODAR (Surface layer NOn-Doppler Acoustic Radar), which is designed to measure the height and turbulence intensity of the atmospheric boundary layer on the Antarctic plateau. This is for example of relevance for astronomers wishing to plan future optical telescopes there. SNODAR works by sending an intense acoustic pulse into the atmosphere and listening for backscatter off inhomogeneities resulting from temperature gradients and wind shear. The theory of operation is very similar to that of the well known underwater sounding techniques of SONAR. SNODARs are monostatic acoustic radars with a minimum sampling height of 5 m, a range of at least 200 m, and a vertical resolution of 1 m. SNODARs operate at frequencies between 4 kHz and 15 kHz. Such high frequencies propagate relatively well in the low temperature of the Antarctic atmosphere (Bonner et al. 2008).

7.2.4 RASS

A radio-acoustic sounding system (RASS) operates acoustic and electro-magnetic sounding simultaneously (Marshall et al. 1972, VDI 3786 part 18 (see Appendix)). This instrument is able to detect the acoustic shock fronts of the acoustic pulses and to determine their propagation speed from the backscattered electro-magnetic waves. This propagation speed is equal to the speed of sound which in turn is a known function of air temperature and humidity. Two different types of RASS have been realised (Engelbart & Bange 2002): a Bragg-RASS and a Doppler-RASS.

A Bragg-RASS is a windprofiler (see Ch. 7.2.2 above) with an additional acoustic emitter (see Figs. 66 and 67). When the Bragg condition is fulfilled, i.e. the wavelength of the sound waves, λ_a is half that of the electro-magnetic waves, λ_e, then there is optimal backscatter of the electro-magnetic waves from the acoustic waves. The electro-magnetic signal is emitted at a fixed frequency, but the emitted sound signal is a chirp signal with varying frequency, f_a. From the sound wave length $\lambda_{a,B}$ at which optimal backscatter occurs the propagation speed of the sound signal can be determined via the following dispersion relation:

$$c_a = \lambda_{a,B} / 2 * f_a. \qquad (7.13)$$

For a VHF windprofiler operating at 50 MHz, a sound frequency of about 100 Hz is used, while for a UHF windprofiler operating at 1 GHz, a sound frequency around 2 kHz is most suitable to fulfill the Bragg condition (Figs. 77 and 78). Because the attenuation of sound waves in the atmosphere is strongly frequency dependent, a UHF RASS can detect temperature profiles up to about 1.5 km height, whereas a VHF RASS can observe temperature profiles throughout the troposphere.

A Doppler-RASS (or SODAR-RASS) is a SODAR with an additional electro-magnetic emitter and receiver (Fig. 76) operating at a frequency $f_{e,0}$. From the Doppler shift, Δf_e (see (7.4)) of the electro-magnetic radiation which is backscattered at the density fluctuations caused by the sound waves, the propagation speed, c_a of the sound waves is determined by.

$$c_a = -0.5 c * \Delta f_e / f_{e,0}) , \qquad (7.14)$$

Fig. 76. SODAR-RASS consisting of a Doppler SODAR (three white antennas in the middle) and two electromagnetic antennas for the measurement of temperature profiles. The upwind antenna (here to the left) emits radio signals, the downwind antenna (here to the right) receives the radio signals. The fences inhibit the access of humans too close to the radio antennas.

Fig. 82. Telescope of an aerosol LIDAR (out of operation).

several kilometres above the instrument. Range determination for the profiles is done via the signal travel time. Two different types of ceilometers are in use: those with two optical axes (one for the emitted light beam and one for the receiving telescope) and those with one optical axis, where the emitted beam is sent through a small hole in the mirror of the receiving telescope. Instruments with one optical axis are able to yield useful information from a distance of about 30 m (Fig. 80), while instruments with two optical axes have an insufficient overlap in the lower 150 m (Fig. 81) so that backscatter information from this height range can not reliably be interpreted (Münkel et al. 2007). A Doppler shift analysis for the determination of the velocity component along the line of side is not made.

7.2.5.3 Differential absorption LIDARs

A DIAL is a special backscatter LIDAR. The basic idea of differential absorption LIDARs (DIALs) is the emission of light pulses at two neighbouring frequencies (Weitkamp 2005). The frequencies are chosen such that one frequency is optimally absorbed by the sought trace gas (λ_{max}), while the other frequency (λ_{min}) experiences insignificant absorption. The two frequencies are chosen so close to each other that no difference in the aerosol backscatter is detectable and the influence of the aerosol can be eliminated by subtracting the two signals from each other. The detected intensity of the backscattered light at frequency, λ_{max} is inversely proportional to the concentration of the absorbing trace gas. The ratio of the backscattered optical intensities at the two frequencies is (Colls 2002)

$$I(\lambda_{max}) / I(\lambda_{min}) = exp(-2rN\sigma), \tag{7.19}$$

where the distance of the backscattering air volume is r, the number concentration of the absorbing trace gas is N, and the absorption cross-section of the absorbing gas is σ. A DIAL yields a vertical resolution of about 75 m for ozone measurements, a temporal resolution of about 60 s for a height range between several hundreds of metres and many kilometres.

The following types of laser are available for a DIAL: a) dye laser with frequency multiplier (the wave lengths may be chosen arbitrarily but these instruments require much attendance and adjustment, so that they are not suitable for monitoring purposes), b) Nd:YAG laser (1064 nm) with frequency multiplier (266 nm) and subsequent Raman shift in deuterium (289 nm) and/or hydrogen (299 nm), c) krypton fluoride excimer laser (248 nm) and subsequent Raman shift in deuterium (268, 292,

7.2 Active sounding methods

319 nm) and/or hydrogen (277, 313 nm). Wulfmeyer (1998) developed an enhanced DIAL based on an alexandrite laser with reduced noise and reduced systematic errors that is able to measure water vapour fluctuations for measurements of the moisture flux.

7.2.5.4 Raman LIDARs

A further possibility for the detection of trace gas profiles in the atmosphere is the operation of a Raman LIDAR, which records radiation from Raman scattering from trace gas molecules (Cooney 1970, Ansmann et al. 1990, Turner et al. 2002, Weitkamp 2005). With normal elastic scattering, a molecule absorbs a photon, enters an excited state, and immediately emits (backscatters) the photon again when returning to its ground state. With inelastic Raman scattering, the molecule does not return to the ground state, but to a neighbouring energy level so that a photon with a slightly changed (usually greater) wavelength is emitted. The observed wavelength shift due to the excitation of rotation or vibration of air molecules is characteristic for a given trace gas species (Fig. 83a). The procedure is named after the Indian physicist Chandrasekhara Venkata Raman who first reported the experimental discovery of this wavelength shift in 1928. The disadvantage of the method is that the inelastically backscattered signal intensities are two to three orders of magnitude lower than with the elastic Rayleigh scattering. Therefore, measurements with a Raman LIDAR are possible only at nighttime or with special provisions to exclude other scattered light.

The scattering air molecules in a relatively cool atmosphere are nearly always in their energetic ground state. Therefore, light from Raman scattering has usually less

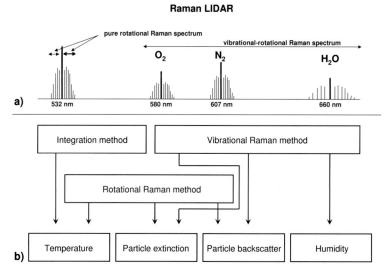

Fig. 83. Schematic of Raman scattering and Raman LIDAR evaluation. a) Example of backscattered wavelengths when light at 532 nm is emitted (simplified from Behrendt et al. 2002). b) Available evaluation methods for Raman LIDAR data to derive temperature, particle (including cloud droplets) property and humidity profiles.

energy than the radiation emitted from the LIDAR (Stokes scattering). With rising temperature, an increasingly number of air molecules can be in excited states, so that Raman scattering by jumps to energy levels below the initial level also becomes possible (Anti-Stokes scattering). Consequently, scattered light with a smaller wavelength than the emitted one can also be detected. But this more energetic radiation from anti-Stokes scattering is weaker than the less energetic radiation from Stokes scattering. Stokes and anti-Stokes lines are seen in Figure 83a directly to the right and the left of the emitted wavelength (pure rotational Raman spectrum).

A Raman LIDAR permits, employing different evaluation methods (Fig. 83b), the detection of trace gas profiles from characteristic wavelength shifts, the determination of type and size of aerosols from extinction measurements, the classification of aerosols from depolarization measurements, and the determination of air temperature from analysing the temperature-dependent ratio of different Raman frequencies. Stratospheric temperature profiles can be derived with the integration method (Keckhut et al. 1990). Temperature and particle profiles in the troposphere are deducible with the rotational Raman method (Behrendt & Reichardt 2000). Humidity and particle profiles in the troposphere are obtained with the vibrational Raman method (Behrendt et al. 2002).

7.2.5.5 Resonance fluorescence LIDARs

Resonance fluorescence occurs at the wavelength of the exciting radiation (the energy of the incident photon equals the energy difference between two energy levels of an ion, a molecule, or an atom). The term is also used to designate radiation emitted by an atom which has the same wavelength as the longest one capable of exciting its fluorescence, e.g., 122.6 nm in the case of the hydrogen atom, and 253.7 nm in the case of the mercury atom (IUPAC Gold book, http://goldbook.iupac.org).

A resonance flourescence LIDAR usually exploits the first case where the emitted laser light is equal to the fluorescent light. Therefore, such a LIDAR is also called resonance backscatter LIDAR (Weitkamp 2005). The backscatter cross-section of resonance fluorescence is 14 orders of magnitude larger than that of Rayleigh backscatter. Resonance fluorescence is thus well suited for the detection of trace substances in the mesosphere and the lower thermosphere at a height range of 75 to 115 km (Chu & Papen 2005). Hence even species with relatively low concentrations have sufficiently high backscatter intensities. Metal ions in the high atmosphere can be particularly well observed by this technique. An evaluation of the Doppler broadening and shift of the sodium D2 line at 588.9 nm allows for a determination of temperature and radial wind speed in the high atmosphere (e.g. Gardner 2004).

A resonance fluorescence LIDAR needs special lasers as a light source. It requires a tunable laser in order to obtain the desired resonance frequencies. The emitted laser light must be narrow-banded (ideally its width is narrower than the width of the resonance line of the trace substance to be detected) and the frequency must be stable for the whole observation time. Wavelengths less than 300 nm cannot be used because the stratospheric ozone absorbs such light (Weitkamp 2005).

7.2 Active sounding methods

7.2.5.6 Doppler wind LIDARs

Analysing the Doppler shift of backscattered laser light (see (7.4)) allows for the determination of wind speed along the line of sight in the same way as from SODAR or RADAR measurements. However, with a Doppler wind LIDAR, the relative frequency shift is much less than with a SODAR. There are two principal techniques used to analyse the Doppler shift of backscattered light: (a) the direct or incoherent determination of the frequency of the backscattered light with a high-resolution spectrometer or with a Fabry-Perot interferometer (Abreu et al. 1992), or (b) by coherent or heterodyne determination (Grund et al. 2001). Coherent or heterodyne implies that the backscattered signal is mixed with a signal from a second (local) oscillator. The resultant beat frequency is much lower and can easily be analysed. While the first direct method is constructionally simpler, it requires the use of narrow-band optical filters through which the radiation must pass. The exact analysis of the beat frequency on the other hand, is quite simple (Weitkamp 2005). Therefore, most wind LIDARs are currently based on the heterodyne method. Conical scanning (Fig. 84) is the usual mode of operation of a wind LIDAR in order to measure vertical profiles of the three-dimensional wind vector.

7.2.5.6.1 Range determination by signal delay

Range-resolved remote sensing systems transmit signals in pulses, which are then scattered by atmospheric inhomogeneities or suspensions (e.g., aerosol, droplets), sending a small fraction of the transmitted energy back to the receiver, as outlined in the previous section. Distance to the measurement volume is determined from the time of flight of the signal pulse. Overviews of state of the art LIDAR techniques for wind and turbulence measurements using signal delay for range determination are given by Hardesty & Darby (2005) and Davies et al. (2003). For the detection of near-surface wind profiles relevant for the operation of wind turbines a special small wind LIDAR using signal delay for distance determination has been designed (Fig. 85).

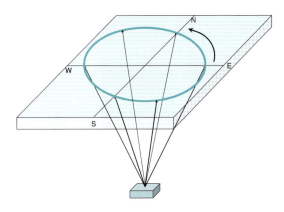

Fig. 84. Conical scanning pattern of a wind LIDAR in order to measure profiles of the three-dimensional wind vector.

Fig. 85. Small pulsed Doppler wind LIDAR for the measurement of wind profiles in a height range between 40 and 200 m. Distance determination is by light pulse travel time (see Fig. 71).

7.2.5.6.2 Range determination by beam focussing

Recently, a transportable continuous-wave wind LIDAR measuring in the near infrared at a wavelength of 1.575 μm (Fig. 86) has been designed and built. It exploits recent developments in optical fibre and related components from the telecommunications industry in order to simplify the alignment and construction of the interferometer that forms the core of the LIDAR. The system emits a continuous-wave (CW) beam, and detection of the wind speed at a given range is achieved by focusing (Fig. 87), rather than by the time-of-flight method of pulsed systems. The system cannot distinguish between air motion towards and away from the LIDAR, and this leads to an ambiguity of 180° in the derived value of wind direction. This is easily resolved, however, by making reference to a simple wind direction measurement at a height of a few metres. The profile of the three-dimensional wind vector is yielded by scanning a cone with a 30° half angle once per second (Banakh et al. 1995, Emeis et al. 2007a, Kindler et al. 2007). Hence the diameter of the measurement volume will be 173 m at a height of 150 m. The probe length increases roughly as the square of the height (Fig. 87). As an example, the vertical resolution is ~± 10 m at a height

Fig. 86. Small continuous-wave Doppler wind LIDAR for the measurement of wind profiles in a height range between 10 and 200 m. Distance determination is by beam focussing (see Fig. 87).

7.2 Active sounding methods 151

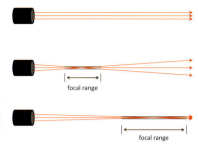

Fig. 87. Schematic of distance determination by beam focussing. Top: unfocussed beam, middle: short distance focussing, below: longer distance focussing. The distance-depending length of the focal range is indicated.

of 100 m. Strong reflections from particles and other moving objects outside the focal range (e.g. due to smoke, fog or birds) can lead to spurious Doppler returns (Harris et al. 2001), but these effects can be recognised and mitigated by signal processing techniques.

7.2.5.6.3 Range determination by optical coherent tomography

An alternative range determination method is the use of optical coherence tomography which has been applied successfully in medical examinations (Fercher et al. 2003). This research (Bennett & Christie 2006, Bennett et al. 2006), which has led only to single case studies but not to a viable method for meteorological purposes so far, uses a fibre-optic based continuous-wave Doppler LIDAR. Infra-red radiation at $\lambda = 1.55$ µm from a distributed feedback laser diode is split between a reference path and a signal path. Radiation following the signal path is amplified and transmitted towards a scattering target, whence some tiny proportion is returned to the instrument. It is then mixed by a coupler with a delayed radiation that has followed the reference path. A narrow-band beat signal can only be expected when the length of the delay line matches the path length in the atmosphere to within the coherence length of the source. Scattering from other distances will give a beat signal widened by twice the linewidth of the source (Harris et al. 1998). The range resolution of the above mentioned system is 34 m, the range is between 42 m and 217 m. For reliable operation, the atmospheric visibility should be below 30 km.

7.2.6 Further LIDAR techniques

Some further techniques for optical velocity measurements exist which up to now have not proven to be suited for wind speed observations in the atmosphere. For instance, also based on aerosol backscatter is laser Doppler anemometry (LDA, Durst et al. 1976), sometimes called laser Doppler velocimetry (LDV), which relies on the interference of two coherent laser beams. Using a beam splitter, the instrument emits two coherent beams which cross each other at some distance (probably much less than 100 m) from the device. At the position of the cross-over, an interference pattern is formed (the frequency of the resulting beats is also called the Doppler frequency which explains the name of this measurement technique). Aerosol particles floating in the air passing through this cross-over region are alternately illuminated or in darkness. From the frequency of this intensity oscillation, the air speed perpendicular to the fringes of the interference pattern can be determined. Both velocity components can be derived with this technique by producing a second pattern turned by 90° with respect to the first one. The method has been developed for wind tunnels. Its maximum range is too small for atmospheric measurements where the ambient light is additionally disturbing. Also not suited for atmospheric purposes are particle imaging velocimetry (PIV, Adrian 1991) and Doppler global velocimetry (DGV, Meyers 1995).

7.3 Active path-averaging methods

The range-resolving methods in Chapter 7.2 have been designed mainly for vertical sounding of the atmosphere. Here, in Chapter 7.3, path-averaging techniques are presented, which are mainly used for measurements along horizontal paths. Using longer paths also allows for the measurement of trace substances with very low concentrations because the number of optically active molecules along the path increases with path length. On the other hand, the received light intensity decreases with increasing path length due to beam dispersion, absorption, and aerosol scattering. The choice of the suitable path length thus is a trade-off between the two effects. Local inhomogeneities along the path are averaged out by the technique. This makes the results from these techniques quite suitable to be compared with results from numerical simulation techniques that usually yield volume averages.

7.3.1 Scintillometers

Scintillometers use a bi-static method that analyses refractive index fluctuations influencing the propagation of optical (e.g. in the near infrared at 940 nm) or electromagnetic waves (e.g., microwaves at 94 GHz having a wave length of 3.2 mm) along the path from the emitter to the receiver. This fluctuation phenomenon, which is called scintillation, is the same which makes stars twinkling at night. Two types of optical scintillometers are available: a small aperture or laser scintillometer (SAS, see, e.g., de Bruin 2002) and a large aperture scintillometer (LAS, see e.g. Kleissl et al. (2008) for a LAS intercomparison study). The first operate with a thin laser beam, the latter with a wide light beam with a diameter of 10 to 20 cm, which is either produced by optical widening of a normal light beam with a diffusor or by a larger array of light emitting diodes (LED). The more modern method with the LEDs makes it easier to produce a uniform light intensity over the whole beam width. The advantage of LAS compared to SAS is that LAS are not so much effected by a satu-

7.3 Active path-averaging methods

ration of the signal along the path (for saturation effects see, e.g., Kohsiek et al. 2006). On the other hand, due to the large beam diameter, LAS in contrast to SAS cannot be used to determine the inner turbulence length scale, which is in the order of 10 mm and is a direct measure for the dissipation rate of kinetic turbulent energy. While SAS and LAS are most sensitive to temperature fluctuations, microwave scintillometer (MWS) also register scintillation from water vapour fluctuations so that the combination of MWS and LAS can be used to derive latent heat fluxes in the surface layer (see, e.g., Beyrich et al. 2005).

The path length can be chosen from between several hundreds of metres to several kilometres, depending on the power of the emitter. The measurement yields the structural parameter C_n^2 (n = refractive index) related to the turbulence intensity of the air. This parameter may be used to estimate the magnitude (but not the direction) of vertical turbulent fluxes (see, e.g., Thiermann & Grassl (1992), or Beyrich et al. (2002)). A scintillometer must be placed in the atmospheric surface layer where Monin-Obukhov similarity theory applies. The scintillometer beam height and set-up distance have to be carefully chosen, such that the instrument operates in the weak scattering regime, where the structure parameter of the refractive index C_n^2 is small enough that it can be derived from the variance of signal intensity I from first-order scattering theory (Clifford et al. 1974).

Some scintillometers, called crosswind scintillometers, operate with two parallel beams in the horizontal (Lawrence et al. 1972). These instruments can estimate the cross-wind speed from the time-lagged cross-variance between the signals of the two beams as well. Together with the knowledge of the wind direction, thus the horizontal wind speed can be derived. Furthermore, if the surface roughness length is available, the surface layer similarity equations for the wind speed and C_n^2 can be solved iteratively to yield two prime turbulence variables: the friction velocity u_* and the surface flux of sensible heat. A sensitivity analysis suggests that, in optimum conditions, a crosswind scintillometer should be capable of providing u_* and the turbulent heat flux with uncertainties of 10 %–15 % and 20 %–30 %, respectively (Andreas 2000).

7.3.2 FTIR

The Fast Fourier Infrared (FTIR) absorption spectroscopy (Bacsik et al. 2004) in its active version is a bi-static method operating a heat source (broad-band infrared radiation source, called a glo(w)bar, see Fig. 88) and a spectrometer based on a Michelson interferometer (Fig. 89). The fundamentals of this spectroscopic method are explained in Smith (1996). Typical path lengths are several hundreds of metres. In a

Fig. 88. Glo(w)bar that serves as radiation (heat) source for FTIR measurements of horizontal path-averaged trace gas concentrations (preferably greenhouse gases).

Fig. 90. DOAS for horizontal path-averaged trace gas concentrations. Lamp and receiving telescope are seen in the foreground, the retro-reflecting mirror in the background. In real operation, the distance between lamp and mirror is several hundreds of metres.

a standard measurement method (Platt & Perner 1984). In contrast to the FTIR spectroscopy, the spectral resolution is lower (f/Δf ~ 100 – 1000, Hase & Fischer (2005)). A complete description of the DOAS method is given in Platt & Stutz (2008). Technical details are described in the VDI guideline 4212 part 1 (see Appendix).

The DOAS light source should ideally emit white light. For this purpose, xenon or halogen lamps are used together with a mirror with 0.3 m diameter and 0.25 m focal length. In special cases, e.g., the detection of OH radicals, dye lasers may be employed as light source. The receiving telescope (focal length 0.25 to 0.85 m) focuses the light on a grate with 600 to 2160 lines per millimetre (Platt & Perner 1984). DOAS detects trace gases which absorb in the UV or visible spectral range (see Tab. 25).

7.3.4 Quantum cascade laser

Quantum cascade lasers are used for active path-averaged trace gas concentration measurements as well (Jiménez et al. 2004). This laser light source (see also Ch. 5.1.1.1) emits very narrow-banded radiation in the infrared spectral range (3 to 5 μm

Table 25. Spectral ranges for DOAS measurements in absorption mode (Platt & Perner 1984).

Species	spectral range (nm)
SO_2	200–230, 290–310
CS_2	200–220, 320–340
NO	215, 226
NO_2	330–500
NO_3	623, 662
HNO_2	330–380
O_3	220–330
CH_2O	250–360
OH	308

and 8 to 13 µm), which allows for a detection of single absorption lines in the infrared. A passive retroreflector is employed to direct the emitted beam back to the receiving unit. During each pulse of about 200 ns, the frequency of the infrared light varies over a certain frequency range due to the warming of the laser during operation. This permits a much more species-specific measurement than with the relatively broad-banded FTIR technology. Rather, interference from unwanted species is negligible. In contrast to the DOAS, quantum cascade lasers are not hampered by atmospheric turbidity, making path lengths of several kilometres possible. The method can, e.g., be employed for the detection of O_3, H_2O and CO_2.

7.4 Passive methods

Passive methods make use of radiation irradiated to the measurement site by natural processes. The methods can be classified into staring methods, methods that integrate over a certain field of view, and imaging methods. Latter methods can again be divided into scanning methods and quasi-photographic methods. Most famous products of the last kind are satellite images.

7.4.1 Radiometers

Radiometers record the intensity and sometimes additionally the polarisation of the incident radiation. This group of instruments also comprises the temperature sensors described in Chapter 3.1.7 and the radiation sensors described in Chapter 6.1. The use of radiometers in remote sensing aims to determine the spectral and physical characteristics of incoming radiation. By an appropriate choice of the detector material the desired spectral sensitivity of the instrument can be achieved. Indium antimonid for instance, is suited to the detection of near infrared radiation, while cadmium tellurid on the other hand is sensitive to longer infrared radiation (Strangeways 2003).

Radiation intensity is a function of the temperature of the emitting object (as described by Planck's law, see (3.14)). On the low frequency side ($h\nu/(kT) >> 1$) the Rayleigh-Jean approximation is valid, where

$$E(\nu) = 2kT \nu^2 / c^2. \tag{7.22}$$

Here, ν is frequency, $k = 1.381 \cdot 10^{-23}$ J K^{-1} is the Boltzmann constant, T is absolute temperature, and c is the speed of light. The so-called brightness temperature, T_b can be obtained from (7.22) by an inversion. For a black body (emissivity $\varepsilon = 1$), brightness temperature equals the true temperature of the object. For lower emissivities, the brightness temperature is lower than the true temperature, where

$$T_b = \varepsilon T. \tag{7.23}$$

7.4.3 Infrared-Interferometer

The Atmospheric Emitted Radiance Interferometer (AERI) is a high spectral resolution interferometer sounder which records spectra in the infrared (3.3 to 18.2 µm) with a spectral resolution of better than 1 cm^{-1}. These radiance spectra are then converted into temperature and humidity profiles by an inversion of the radiation transfer equation (Knuteson et al. 2004 a, b). This inversion requires a first guess which may be taken from a radiosonde measurement or from climatology. The instrument contains two calibration blackbodies, one at 60 °C and one at ambient temperature which allow for a self-calibration of the instrument. AERI can measure up to a height of about 3 km above ground. The vertical resolution is 100 to 200 m.

7.5 Tomography

Tomographic methods are able to deliver cross-sections through an air volume. It is often also called computed tomography (CT). The optical method is based on integrated optical measurements along a net of intersecting paths covering the area or volume of interest and a subsequent inverse computation of the absorption along the cross-sections. The optical emitters and receivers can be installed either along the perimeter of the area of interest (if the area to be probed should not be accessed) or the light source can also be erected inside the area with all the detectors at the perimeter. The latter mode of installation can be realised with less effort (Hashmonay et al. 1999, Wu et al. 1999). The original idea for this technique dates back to Byer & Shepp (1979). Tomographic methods are also applied in several other disciplines such as medicine, geophysics, or oceanography.

The acoustic travel time tomography is based on an analogue principle (Wilson & Thomson 1994, Raabe et al. 2001, Tetzlaff et al. 2002). Here, acoustic emitters and receivers are mounted on the perimeter of an area (Fig. 92). From the travel time of

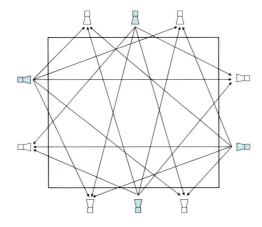

Fig. 92. Possible arrangement of four loudspeakers (full symbols) and six microphones (open symbols) for acoustic tomography for the measurement of the horizontal distribution of wind speed and temperature.

the sound pulses, the areal distribution of acoustic air temperature (see (3.13)) can be obtained. If the sound pulses are sent back and forth on all paths the two wind components in the plane enclosed by the instruments can also be obtained. In this, tomography can be operated in analogue to a sonic anemometer.

Generally, stationary conditions are required until all measurements entering one inversion are taken. Optical tomography is obstructed by fog and strong precipitation, acoustic tomography by strong ambient noise and by heavy precipitation. Orographic features and buildings may prevent a sufficient coverage of the measurement area. Three evaluation techniques for tomographic measurements are sketched below.

7.5.1 Simultaneous Iterative Reconstruction Technique (SIRT)

For the SIRT algorithm (Humphreys & Clayton 1988, Trampert & Leveque 1990, Tetzlaff et al. 2002), the area is subdivided into regular smaller areas $j = 1, J$, which are covered by the paths $i = 1, I$. Parts l_{ij} of the paths, i cross the respective subareas j. For the travel times τ and the respective inverse velocities, s, the following relation holds:

$$\tau_i = \sum_{j=1}^{J} l_{ij} s_j \qquad (7.28)$$

Because we get the travel times from measurements, the inverse of (7.28) must be solved. The necessary first guess for the inverse velocities, s_j are obtained by a simple back projection of the measured travel times, τ_i on the subareas. Then (7.28) is solved and the differences between the measurement and the solution are determined. These differences are then once again distributed by back projection on the subareas and added to the original inverse velocities. The iteration is repeated until a given accuracy, $\Delta \tau_i$ is reached.

7.5.2 Algebraic Reconstruction Technique (ART)

For the ART algorithm the area is also subdivided in regular subareas (pixels). The initial values for all pixels are set to the mean value over all measurement pathes. Then the algorithm adjusts the values for all those pixels which are crossed by measurement paths in such a way that the mean square difference between the values of pixels and paths is minimized. This algorithm is repeated until the deviations are lower than a predefined value. Usually, there are more pixels than paths so that the algorithm is under-determined. The developers of the technique (Todd & Leith 1990, Todd & Ramachandran 1994) have shown that for favoured conditions, reconstructions of the field distribution over the area can be obtained up to a ratio between pixels and paths of 6.7 to 1. The principal problem of underdetermination can be solved either by more paths (this means higher measurement efforts) or by a subdivision into less pixels (i.e., a coarser spatial resolution).

7.5.3 Smooth Basis Function Minimization (SBFM)

Reacting to the problems with ART, Drescher et al. (1996) developed an alternative algorithm called Smooth Basis Function Minimization (SBFM). SBFM is not based on a subdivision of the area but the spatial distribution is approximated by overlaying several bivariate Gauß functions. Each of these Gauß functions is determined by six parameters: the x and y coordinate of the maximum, the two standard deviations along the main axes, the height of the maximum and the angle between the main axes and the coordinate system. Two or three Gauß functions are used for the approximation. SBFM is based on the Amebsa routine in the Numerical Recipes (Press et al. 1992). It employs a combination of the "Simplex" method (Ch. 10.8 in Press et al. 1992) with a simulated annealing technique (Ch. 10.9 in the same book) in order to find the parameters for the Gauß functions which lead to a minimization of the mean square deviation. Annealing is a technique which is able to find the global minimum of a function even if there are secondary minima close to the first guess. The algorithm is initiated with a constant mean value, then it is applied to one Gauß function, then to two functions, and so on until the mean square deviation falls below a predefined threshold. SFBM delivers considerably smoother distributions than ART. The method is rather time-consuming and the quality of the first guess is decisive. ART on the other hand is rather fast and could be used to produce a first guess for SBFM (Drescher et al. 1996).

8 Remote sensing of atmospheric state variables

This and the following four Chapters 9 to 12 serve to list available remote sensing methods for the most important variables in atmospheric monitoring and research. The methods are usually relying on inversion methods (see Ch. 2.3, 7.1, and 7.5).

8.1 Temperature

The temperature of gases, aerosols, and water droplets in the air can be remotely determined with passive techniques either from their emission of infrared, thermal, or microwave radiation. Air temperature can alternatively be retrieved with active techniques from a measurement of the speed of sound or from a determination of light or radio wave absorption.

8.1.1 Near-surface temperatures

Near-surface air temperature may be obtained as the lowest value from profiling radiometer measurements within the boundary layer if sufficiently low range gates are available (see Ch. 8.1.2.1 below). Path-averaged near-surface air temperatures may be obtained from acoustic measurements which determine the travel time of a sound pulse (see Ch. 7.5). This travel time can be converted into the speed of sound and further into the acoustic temperature, T_a (see (3.13)) that is close to the virtual temperature, T_v (see (3.5)).

The direct detection of the near-surface temperature, T_{air} from satellites or high-flying aircraft over land surfaces is very difficult because the emissivity of the land surface is close to unity. Over ocean surfaces the retrieval is somewhat simpler due to the lower emissivity of this surface (English 1999). A microwave method has been developed (Basist et al. 1998) to calculate near-surface temperature using measurements from the Special Sensor Microwave/Imager (SSM/I).

Indirect determination of T_{air} from satellites is possible from skin surface temperature, T_s and known surface properties such as the vegetation index. The common satellite-based techniques for estimating T_{air} rely on the negative relationship between a vegetation index (often the normalized difference vegetation index (NDVI, for a definition see Ch. 11.1.4)) and a calculated T_s from thermal channel data. A relationship is defined through regression analysis, and extrapolations are made up to some theoretical vegetation index value that represents a full canopy as viewed by the sensor (Prihodko & Goward 1997, Riddering & Queen 2006).

8.1.1.1 Small-scale horizontal temperature gradients

The acoustic tomography technique (Ch. 7.5) allows for the determination of the structure of the horizontal temperature distribution in an area enclosed by sound sensors and receivers. Such areas can have horizontal extensions of hundreds of metres. The spatial resolution depends on the number of pathes crossing the area of investigation.

8.1.2 Temperature profiles

The knowledge of vertical profiles of temperature and moisture are necessary for the assessment of the static stability of the atmosphere. A basic overview of temperature and water vapour profile retrievals is given in Westwater (1997) and a tutorial for microwave detection of these profiles by Westwater et al. (2005a). A more detailed overview of microwave profiling with instrument descriptions can be found in Westwater et al. (2005b).

8.1.2.1 Temperature profiles in the atmospheric boundary layer

Passive microwave radiometers (Ch. 7.4.1) are able to deliver vertical temperature profiles up to about 600 m above ground with a vertical resolution of about 50 m (Solheim et al. 1998, Westwater et al. 1999). A scanning instrument originally developed in Russia (Troitsky et al. 1993, Kadygrov & Pick 1998) detects the emission from the oxygen band at 60 GHz at eleven elevation angles between the horizontal and the vertical. A subsequent inversion procedure yields the temperature with an accuracy of 0.5 K. Martin et al. (2006b) describe an instrument, which scans the upper hemisphere over the instrument, and which can yield the temperature up to 5 km above the ground. The required inversion procedure needs first guess profiles taken from a radiosonde climatology.

Passive infrared radiometers (e.g., the Atmospheric Emitted Radiance Interferometer AERI, Ch. 7.4.3) deliver temperature profiles up to 3 km height with a vertical resolution of 100 to 200 m (Smith et al. 1999, Feltz et al. 2003). The evaluation of the radiometer data requires a first guess which may be taken from radiosonde measurements or from climatology. The accuracy is roughly 1 K for temperature profiles.

Apart from the just mentioned techniques and the FTIR method (see Ch. 8.1.2.2 below), no further single instrument techniques for the retrieval of the atmospheric temperature profiles are available. An active two-instrument acoustic-electromagnetic technique to obtain temperature profiles is the RASS method (Ch. 7.2.4), which essentially determines a vertical profile of the speed of sound which is converted

8.1 Temperature

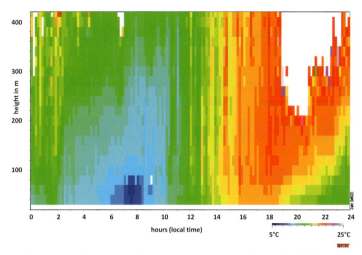

Fig. 93. Time-height cross-section of potential temperature captured by a RASS. Blue colours denote cold temperature, red colours warm temperature. On the left-hand side a stable nocturnal stratification is depicted (warmer lighter air over colder heavier air). On the right-hand side a daytime convective is depicted (well-mixed, constant potential temperature with height). The missing data (white area) to the upper right is due to too high wind speeds (above about 15 m/s) which move the sound pulses of the RASS out of the focus of the radar beam.

into a temperature profile. With Doppler-RASS, profiles over several hundreds of metres are obtainable (see Fig. 93 as an example), while with Bragg-RASS profiles up to a few kilometres height can be obtained.

8.1.2.2 Temperature profiles in the troposphere and stratosphere

Surface-based multi-frequency microwave radiometers (Ch. 7.4.1) can be employed to obtain the temperature profile up to 3 km height with a vertical resolution of about 500 m (Strangeways 2003, Dabberdt et al.2004) or even up to 10 km height (Engelbart 2005). The temperature is deduced from the radiation emitted at the oxygen band between 50 and 60 GHz. Because the mixing ratio of oxygen is constant, this radiation is merely a function of temperature. Measurements near the centre of the oxygen band give information on the temperature near the instrument because at this frequency, the atmosphere is rather opaque. With increasing distance from the band centre, the atmosphere becomes more transparent and the temperature information is from increasingly further distances away. The attribution of the distance to the emitting volume is done from the weighting function for the chosen frequency via an inversion of the radiation transport equation (Solheim & Godwin 1998, Güldner & Spänkuch 2001).

Infrared radiometers with rotating filter wheels (Ch. 7.4.1) containing about 20 different filters allow for satellite observations in several infrared spectral bands within short time intervals. By solving the radiation transfer equation and height attribution from frequency-dependent weight functions which relate radiation emissions to certain heights, a vertical temperature throughout the whole atmosphere can be computed (see e.g. Menzel et al. 1998). The technique is an analogue of the microwave technique described before.

Temperature profiles from the lower troposphere can also be observed with infrared interferometry. The German Weather Service operates an experimental version of such a FTIR spectrometer (Ch. 7.3.2) at the Richard-Assmann oberservatory Lindenberg called EISAR (Emission Infrared Spectrometer for Atmospheric Research, Spänkuch et al. 1996, Engelbart 2005). This passive instrument records radiation from the frequency band between 3.3 and 20 µm.

Alternatively and with a higher vertical resolution vertical temperature profiles can be obtained from surface-based active soundings with a Bragg-RASS (Ch. 7.2.4). This instrument which simultaneously gives vertical wind profiles is increasingly used to fill gaps in the radiosonde network or even to replace radiosonde stations. The German Weather Service is presently (2009) operating four such instruments in Lindenberg (east of Berlin), Bayreuth (Northern Bavaria), Nordholz (Northern Germany, south of the Elbe estuary), and Ziegendorf (Eastern Germany, close to the Baltic Sea coast). The data are used regularly in the routine weather forecast models.

For the determination of the temperature distribution in the stratosphere, the GPS radio occultation method is currently in use. It is applied for a height range of 10 to 35 km above ground; an extension of the measurements to heights up to 120 km seems possible from a physical point of view. The method analyses the influence of the atmosphere on GPS signals with special satellites. Measurements are taken when the satellites cross the horizon (limb sounding). From that, profiles of temperature, moisture, and refractivity are derived from alterations to the signal (Wickert & Schmidt 2005, Wickert & Gendt 2006, Kursinski et al. 1997). Since May 2001, e.g., the German research satellite CHAMP (CHAllenging Minisatellite Payload) has observed 150 to 200 occultations per day. This allows for a global observation of vertical profiles with a vertical resolution of 0.5 km.

DIAL measurements (Ch. 7.2.5.3) can also detect the temperature of a gas in the troposphere (Kalshoven et al. 1981). If the concentration of a selected gas (e.g. oxygen) is known, a determination of the absorption cross-section of a rotational band is possible. The temperature dependence of the absorption cross-section within this band is known from laboratory experiments. Likewise, a Raman-LIDAR (Ch. 7.2.5.4) can also be employed (Arshinov et al., 1983).

8.1.2.3 Temperature profiles in the upper atmosphere

With a resonance fluorescence LIDAR (Ch. 7.2.5.5), the temperature profiles in heights between 75 and about 115 km can be determined from the Doppler-broadening of selected emission lines of metal ions (She et al. 1992).

8.2 Gaseous humidity

Atmospheric humidity is an important part of the hydrological cycle. It is a prerequisite for the formation of clouds and precipitation. The vertical distribution of humidity may have considerable influence on the static stability of the atmosphere. The gaseous moisture content of the atmosphere can be remotely determined with passive techniques either from microwave emissions or from infrared emissions. Active remote sensing is possible from the detection of light absorption.

8.2.1 Integral water vapour content

The integral water vapour content (IWV) of an atmospheric column may be obtained from surface-based passive microwave radiometer measurements (Ch. 7.4.1) at two frequencies if it is not raining. The accuracy is about 1 to 2 mm (expressed in terms of precipitable water, Dabberdt et al. 2004) or even 0.41 kg/m² (Martin et al. (2006b). The latter reports the following relations for soundings at 23.6 and 31.5 GHZ:

$$IWV = 0.0646 + 219.5029\ \tau_{23.6\ GHz} - 130.5823\ \tau_{31.5GHz} \qquad (8.1)$$

or for soundings at the three frequencies 23.6, 31.5 and 151 GHz:

$$IWV = 0.5848 + 211.0855\ \tau_{23.6\ GHz} - 143.6338\ \tau_{31.5GHz} + 2.1612\ \tau_{151GHz}. \qquad (8.2)$$

8.2.2 Vertical profiles

Microwave radiometers (Ch. 7.4.1) may be employed to observe vertical humidity profiles from the surface up to about 10 km height. For this purpose, observations are made in the range of the water vapour absorption line centred at 22.325 GHz. Here, the absorption characteristics in the centre of this line and at its flanks differ (Solheim & Godwin 1998, Güldner & Spänkuch 2001, Engelbart 2005, Martin et al. 2006b). This frequency range is not suited for satellite observations because the atmosphere is quite transparent at these frequencies and emission from the Earth's surface disturbs it too much. Therefore, satellite observations close to 183 GHz are used where the atmosphere is much more opaque. Due to the opacity, satellite observations of the lower troposphere are not possible in this frequency range (Solheim & Godwin 1998).

Infrared interferometers (AERI, Ch. 7.4.3) deliver humidity profiles from heights up to about 3 km above the ground with a vertical resolution of 100 to 200 m (Smith et al. 1999, Feltz et al. 2003). They require a first guess of profiles in the determination of humidity from a radiosonde measurements or from climatology. The time resolution is 10 min and 1 g kg^{-1} for the humidity.

The derivation of the vertical humidity profile is also possible from a pair of satellite-observed temperature profiles obtained from a solution of the radiation transfer equation with assumed vertical profiles of carbon dioxide and of water vapour in a spectral range where carbon dioxide as well as water vapour emit radiation. Because the concentration of carbon dioxide is fairly constant and well known, differences between the two temperature profiles must be due to deviations from the assumed temperature profile. From these deviations, the true water vapour profile can be inferred.

Similar to temperature profiles humidity profiles in the troposphere may also be derived from the retardation of GPS signals (Ch. 8.1.2.2) if at least two different frequencies are recorded. Thus, regional-scale horizontal and vertical humidity distributions may be detected with a network of GPS receivers receiving radiation under different zenith angles (Bevis et al. 1994, Bender et al. 2008). The primarily detected variable is the so-called zenith wet delay (ZWD) or slant wet delay (SWD) in millimetres. This prolongation of a travel path of a radio wave in the moist atmosphere is caused by temperature- and humidity-dependent refractivity, n for radio waves in the air (Smith & Weintraub 1953). The following relation holds (Bender et al. 2008):

$$(n-1) \cdot 10^6 = 77.6 \frac{p_d}{T} + 70.4 \frac{e}{T} + 3.739 \cdot 10^5 \frac{e}{T^2}, \tag{8.3}$$

here the pressure of dry air is p_d and the water vapour pressure is e, both given in hPa and the temperature, T in K.

Humidity profiles in the upper troposphere and the lower stratosphere may be obtained from surface-based measurements utilizing a water vapour LIDAR (a type of DIAL, Ch. 7.2.5.3, Trickl & Vogelmann 2004). This LIDAR emits infrared light pulses at two different wavelengths which are absorbed differently by atmospheric water vapour. In order to avoid large absorption due to high water vapour concentrations in the atmospheric boundary layer, a measuring site high up in the mountains is ideal. At night, the water vapour profile may also be obtained from measurements with a Raman LIDAR (Ch. 7.2.5.4, Engelbart 2005). In contrast to passively operating microwave radiometers, actively operating LIDARs exhibit a much better vertical resolution (about 50 to 100 m). Their drawback is that they can only be used in cloud-free situations. Clouds do not affect microwave radiometers. The determination of the absolute amount of water vapour with a Raman LIDAR is possible only if additional measurements from a radiosonde or a microwave radiometer are available for calibration.

8.2.3 Large-scale humidity distribution

Satellite images taken in the water vapour band (channel WV, see Tab. 21) show the large-scale humidity distribution in the upper troposphere (Fig. 94 lower frame). As the radiation intensity in this frequency band is also a function of temperature, high

8.2 Gaseous humidity

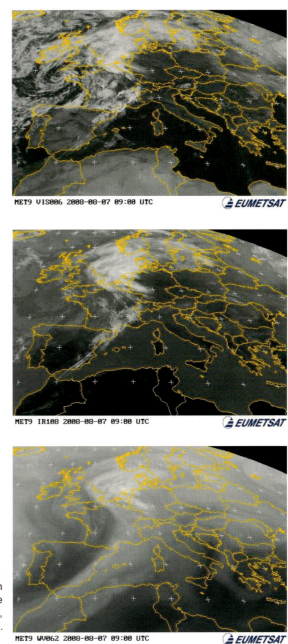

Fig. 94. Satellite images of Europe on Aug. 7, 2008, 0900 GMT. Top: visible (0.6 μm), middle: infrared (10.8 μm), below: water vapour channel (6.2 μm). Copyright 2009 EUMETSAT.

intensities are emitted from relatively warm (i.e. lower) layers in the middle of the troposphere and low intensities are emitted from relatively cold (i.e. higher) layers in the upper troposphere. In any case the atmosphere is optically opaque in this frequency range so that the lower troposphere or the Earth's surface do not contribute to the signal received by the satellite. The measured radiation stems from warmer

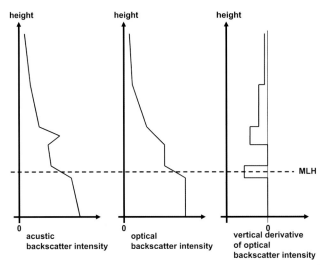

Fig. 96. Principle of mixing height determination from acoustic backscatter intensity (e.g., from a SODAR, left) and from optical backscatter intensity (e.g., from a ceilometer, middle and right).

an automated procedure: the gradient method (Fig. 96), the inflection point method, and the logarithmic gradient method. The gradient method (Hayden et al. 1997, Flamant et al. 1997, for ceilometers see: Emeis et al. 2007b) looks for the strongest decrease of the optical backscatter intensity:

$$MLH_{GM} = min(\partial RSCS/\partial r). \tag{8.11}$$

The inflection point method (Menut et al. 1999) searches for the strongest increase of the vertical decrease of the backscatter intensity:

$$MLH_{IPM} = min(\partial^2 RSCS/\partial r^2). \tag{8.12}$$

The logarithmic gradient method looks for the strongest vertical decrease of the logarithm of the backscatter intensity (Senff et al. 1996):

$$MLH_{LGM} = min(\partial lnRSCS/\partial lnr). \tag{8.13}$$

All three methods deliver slightly different values of MLH, where usually MLH_{IPM} < MLH_{GM}, and where MLH_{LGM} frequently yields the highest value. Following Sicard et al. (2006), MLH determined from the IPM-method is closest to MLH determined via the Richardson method from radiosonde ascents.

8.4.2 SODAR

Overviews of methods to derive MLH from SODAR data are given by Beyrich (1997) and Asimakopoulos et al. (2004). Some additional estimation algorithms for the determination of the height of the convective boundary layer, which is usually outside the vertical range of SODAR measurements, are described in Beyrich (1995). These latter methods rely on extrapolations, similarity theories, and statistical relations. The following sketches an enhanced 'acoustic received echo' (ARE) method combining information from the acoustic backscatter intensity $R(z)$ (measured in dB) and from the variance of the vertical velocity component σ_w (Emeis & Türk 2004). A first height criterion, H_1 of this method looks for rapidly decreasing turbulence with height (left frame in Fig. 96), which frequently signifies the top of nocturnal mixing layers and sometimes indicates the top of convective mixing layers in the absence of temperature inversions is defined as

$$H_1 = z, \text{ if } (R(z) < 88 \text{ dB and } R(z+1) < 86 \text{ dB and } R(z+2) < 84 \text{ dB}). \quad (8.14)$$

The utilization of information from three adjacent layers in this criterion shell prevent the misinterpretation of single outliers. A second criterion, H_2 searches for surface and elevated inversions. Lifted inversions are assumed to be present when secondary maxima of the backscatter intensity occur:

$$H_2 = z, \text{ if } (\partial R/\partial z(z+1) < 0 \text{ and } \partial R/\partial z(z-1) > 0 \text{ and } \sigma_w < 0.70 \text{ ms}^{-1}). \quad (8.15)$$

The additional upper threshold for the variance of the vertical velocity component eliminates the detection of backscatter maxima due to high, thermally induced tur-

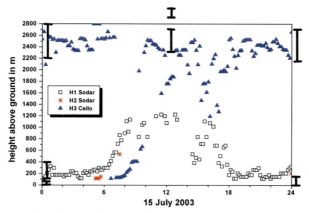

Fig. 97. Example of a mixing-layer height determination from a simultaneous operation of a ceilometer (triangles) and a SODAR (squares and asterisks) for 24 hours from midnight to midnight over a height range of 2800 m. Shallow stable nocturnal surface layers were observed in the early morning and the late evening, while a convective boundary was seen during daytime. At night a residual layer is found between the stable surface layer below and the overall top of the planetary boundary layer. Vertical bars indicate inversions from simultaneous radiosonde observations.

bulence. Surface inversions are connected to high, near-surface backscatter intensities and very low variances of the vertical velocity component:

$$H_2 = z, \text{ if } (R(z-1) > 105 \text{ dB and } R(z) < 105 \text{ dB and } \sigma_w < 0.3 \text{ ms}^{-1}). \tag{8.16}$$

High, near-surface backscatter together with high variances, σ_w would be characteristic for adiabatic or even super-adiabatic surface layers due to strong insolation. The MLH is taken as the lowest value from (8.14) to (8.16). More details on this search algorithm can be found in Emeis & Türk (2004) and an extention of the algorithm to more than one inversion in Emeis et al. (2007b). The simultaneous operation of a ceilometer (Ch. 7.2.5.2) may allow for more detailed analyses of the vertical structure of the boundary layer (especially the direct capture of the top of convective mixing layers and the existence of nocturnal residual layers, see Emeis & Schäfer (2006) and Fig. 97).

A special SODAR for the monitoring of the Antarctic boundary layer height, a SNODAR (see Ch. 7.2.3.1), has been developed recently (Bonner et al. 2008).

8.5 Turbulent fluxes

Remote sensing of turbulent fluxes of an atmospheric property requires the simultaneous observation of wind fluctuations and property fluctuations with high temporal resolution (10 Hz) in the same air volume (see also Ch. 2.1.5 and 6.2 above) with two separate remote sensing devices. An overview of remote momentum and heat flux measurements in the boundary layer and the free atmosphere is given in Engelbart et al. (2007). Moisture fluxes can, for example, be observed by the simultaneous operation of a Bragg-RASS for the wind fluctuations and a water vapour DIAL for the humidity fluctuations (Wulfmeyer 1999a, 1999b) or by the operation of a Doppler LIDAR for the wind fluctuations and a DIAL for the humidity fluctuations (Linné et al. 2007).

Similarity laws for the atmospheric boundary layer may be used to determine the near-surface turbulent fluxes of sensible (H) and latent heat (LE) from surface-layer turbulence measurements with scintillometers (see Ch. 8.3.1). For unstable conditions, Kohsiek (1982) gives

$$H = b \, (C_T^2)^{3/4}, \tag{8.17}$$

$$LE = 1.09 \, b \, (C_T^2)^{1/4}(C_q^2)^{1/2}, \tag{8.18}$$

$$b = 0.55 \, z \, (g/T)^{1/2}. \tag{8.19}$$

For large Bowen ratios when the humidity influence can be neglected, scintillometer measurements at one wavelength are sufficient and the heat flux can be derived from a relation between C_T^2 and C_n^2 from Wesely (1976) and from Monin-Obukhov similarity theory (Kleissl et al. 2008). A comparison of the structure parameters for temperature and humidity from scintillometer measurements with simultaneous eddy-covariance measurements is described in Beyrich et al. (2005) showing a consistent behaviour in time and deviations in the order of 20 to 35 %.

8.6 Ionospheric electron densities

Ionospheric electron densities may be observed with a special RADAR. Air molecules are partly ionized by the incoming short-wave radiation above about 60 to 70 km height. The free electrons make this air electrically conductive. The refractivity, n of such ionized layers for radio waves of frequency, f is proportional to the plasma frequency, f_p where

$$n^2 = 1 - f_p^2/f^2. \tag{8.20}$$

Total reflection occurs for $n = 0$ (i.e. $f = f_p$). Receiving the backscattered intensities of emitted radio waves from 1 MHz to 15 MHz (wavelength 20 to 300 m) in steps of 50 kHz leads to a so-called ionogram. An ionogram is a frequency-height diagram from which the vertical distribution of the electron density may be determined. This technique has been used for more than 75 years now in a global ionosondes network (Jakowski 2005).

8.7 Recommendations for remote sensing of state variables

The continuous monitoring of vertical profiles of temperature, humidity, and wind in the boundary layer and the lower troposphere at given localities is best performed by surface-based remote sensing. The observation of larger areas in the middle and upper troposphere and in higher atmospheric layers is most suitably done from satellites. Geostationary satellites such as METEOSAT monitor the diurnal variation of a given part of the Earth surface (nearly one hemisphere). Sun-synchronized satellites on polar orbits observe different stripes (swaths) of the surface always at the same local time (i.e. the inclination of the satellite's orbit to the sun is kept constant). Satellites at lower heights on non-polar orbits such as SPOT or Landsat offer the best spatial resolution but fly over given areas only in larger time intervals (sometimes an area is not revisited for two or three weeks).

Comparing available surface-based remote sensing technologies, acoustic techniques are the simplest and cheapest option. Their advantage is the relatively high vertical resolution due to the low propagation speed of sound waves. Their greatest disadvantages are the low maximum range and disturbances by ambient noise. Their sound pulses annoy people working or living nearby (in distances up to several hundreds of metres). This restricts their application to the nocturnal boundary layer or the lower part of the daytime boundary layer. Optical and electromagnetic techniques require much more effort and are much more expensive. Their greatest advantage is a maximum range that easily allows for monitoring of the stratosphere and higher layers of the atmosphere.

A RASS allows for the simultaneous observation of wind and temperature profiles. The range and quality of the observations differs for the two variables. Temperature profile detection has higher data availability for low wind speeds. At higher wind speeds, the sound pulse which is to be followed by the electromagnetic waves is advected away from the focal point of the receiving electromagnetic antenna. Wind observations exhibit a higher data availability at larger wind speeds because the turbulence intensity and thus the small-scale temperature fluctuations are usually higher. The drift of acoustic signals with wind speed is not very decisive for the acoustic backscatter.

All active remote sensing techniques which emit focussed radiation, require careful planning in order to avoid disturbances or even damages to living creatures and human beings. The direct access to these instruments must be prohibited or the instruments must be housed so that an interaction with these instruments is possible for authorized persons only. SODAR and RASS must be operated at distances of at least 500 m from housing areas and offices. The pulses must not be emitted in horizontal or near horizontal directions because they can damage eardrums. Maintenance work is permitted only if the device is switched off or ear protection is worn. For optical instruments, eye safety is an important issue. Eye safety may be obtained by restricting the amount of emitted energy, by choosing wavelengths longer than the visible light, or by widening focussed light beams. If none of these precautions are realizable, the instruments must be switched off when persons come close to the light beams. This also includes the approach of aircraft if the instruments are operated vertically. The emission of stronger electromagnetic radiation may also be harmful. RADAR instruments, windprofilers, and RASS instruments must be shielded by fences or other means to prevent individuals from approaching. Fences also serve to prevent interference with other communication techniques operating at similar wavelengths. The operation of active electromagnetic remote sensing instruments (including RASS) often needs a permit from the responsible authorities.

The success of remote sensing depends decisively on obtaining good signal-to-noise ratios. Therefore, ambient disturbances of similar wavelength must be avoided. Acoustic devices should be operated away from stronger noise sources including motorways and railways. Due to rather strong side lobes which unavoidably are emitted from acoustic instruments, the proximity to reflecting objects (houses, trees, electric cables) is a nuisance. For vertical operating instruments, reflecting objects should be lower than about 30° above the horizon around the instrument. Electromagnetic instruments must be shielded to prevent interference with ambient electro-

8.7 Recommendations for remote sensing of state variables

magnetic radiation, e.g. from television broadcasting. The lowest ambient interferences can be expected for the operation of optical instruments.

Many remote sensing techniques deliver volume-averaged values. This makes data from these instruments, in contrast to data from most in-situ instruments described in Chapter 3, well suited to be compared to the results of numerical simulation models which also produce volume-averaged information.

Technical rules for surface-based remote sensing of atmospheric state variables may be found e.g. in VDI (VDI 3786 part 11 (SODAR), part 14 (Doppler-Wind-LIDAR), part 17 (Windprofiler), and part 18 (RASS) and ANSI and CEN guidelines (see Appendix).

Measuring Change in a Changing World®

CO_2/H_2O Monitoring

Photosynthesis

Light Measurement

Methane

Leaf Area

Soil Respiration

GHG Analysis

Eddy Covariance

www.licor.com
1-402-467-3576 (USA)
EnvSales@licor.com

Biosciences

9 Remote sensing of water and ice

9.1 Precipitation

9.1.1 RADAR

Cloud droplets and falling precipitation may be detected by S- to W-band RADARs operating at 2 to 100 GHz (see Ch. 7.2.1). Water droplets and ice crystals interact with radio waves by Rayleigh scattering. The observed RADAR reflectivity Z (see (7.5)) depends on the 6th power of the droplet (or crystal) diameters, D_i in the volume, ΔV:

$$Z = 1/(\Delta V) \, \Sigma \, D_i^6. \tag{9.1}$$

If the droplet spectrum $N(D)$ is known, the sum (9.1) can be formulated as an integral by

$$Z = \int N(D) \, D^6 \, dD. \tag{9.2}$$

The droplet spectrum may be given by the Marshall-Palmer (1948) distribution, which describes the spectrum as a function of droplet diameter, D in mm and precipitation rate, R in mm h^{-1}:

$$N(D) = N_0 \exp(-\Lambda D), \tag{9.3}$$

with $N_0 = 8 \cdot 10^3$ m^{-3} mm^{-1} and $\Lambda = 4.1 \, R^{-0.21}$ mm^{-1}. Inserting (9.3) into (9.2) yields

$$Z = N_0 \, 6! \, 4.1^{-7} \, R^{1.47} = a \, R^b. \tag{9.4}$$

Eq. (9.4) is known as the Z-R relation (Marshall & Palmer 1948) that gives an empirical relation between the RADAR reflectivity and the precipitation rate. The two constants a and b in (9.4) depend on the type of precipitation. Marshall & Palmer (1948) found $a = 296$ and $b = 1.47$. Inserting (9.4) into the RADAR equation (7.1) gives a relation between the intensity of the backscattered radio waves and the precipitation rate, that permits a quantitative precipitation measurement by RADAR. The RADAR reflectivity in (9.4) can vary over several orders of magnitudes. Therefore, the logarithmic measure,

$$dBZ = 10 \cdot \log_{10}(Z/Z_0) \tag{9.5}$$

is frequently used (with $Z_0 = 1$ mm^{6}m^{-3}).

9.1 Precipitation

Table 27. Typical RADAR reflectivities.

Z [mm⁶m⁻³]	dBZ	rain rate [mm h⁻¹]	remarks
1	0	0.04	one drop per m³ with $D = 1$ mm
200	23	1.00	light rain
15 625	42	15	one drop per m³ with $D = 5$ mm
257 800	54	100	heavy rain
10^6	60	205	very heavy rain (most likely with hail)
10^7	70		large hailstones

The appropriate Z-R relation (9.4) must be determined from a comparison of RADAR reflectivities with measured drop spectra (e.g., obtained from disdrometer measurements in the area covered by the RADAR beam). Typical values for uniform rain from stratus clouds are $a = 200$ and $b = 1.6$. For rain from convective clouds, $a = 350$ and $b = 1.4$ is a better choice. For snow, $a = 2000$ and $b = 2$ are typical values. Some values for Z and dBZ and the rain rate R using $a = 269$ and $b = 1.47$ are listed in Table 27.

A weather RADAR for rain detection is usually operated in one of three possible scanning modes (Fig. 98). Rotation around the vertical axis while emitting the beam at a constant low elevation angle leads to circular maps of the precipitation distribution with the position of the RADAR in the centre of these maps. This mode is known as PPI (plane position indicator). Because the RADAR beam is usually bent less than the curvature of the Earth's surface, the height of the beam above the ground increases with distance from the origin in a PPI mode. An elevation angle of 0.5° leads to a height of the beam above ground of 4 km at a distance of 200 km. Successive scans with varying elevation angles can be used to compute circular maps which artificially show backscatter echoes from a fixed height above ground (CAPPI, con-

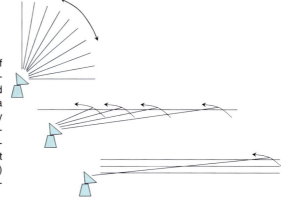

Fig. 98. Modes of RADAR observation of clouds and precipitation. Upper left: Capture of a vertical cross-section for a fixed azimuth angle (RHI), middle: capture of a horizontal distribution in one height level by circular sounding at different elevation angles (CAPPI), below: capture of distance-dependent horizontal distribution (height above ground is increasing with distance) by circular sounding at a fixed elevation angle (PPI).

K- and W-band RADARs (Ch. 7.2.1) employ the reflection of radio waves at cloud droplets. Cloud RADARs at 35 GHz (the MIRA-36 of the German Weather Service in Lindenberg operates at 35.5 GHz, Engelbart (2005)) reach heights up to about 15 km. The vertical resolution depends on the pulse length which range from 100 ns (15 m resolution) to 400 ns (60 m resolution), with a minimum range of 150 m. If signals in two different polarizations are emitted, the determination of cloud microphysical parameters such as the drop size distribution, the liquid water content, and the ice content can be derived (Engelbart 2005).

Two older methods are the use of searchlight and the optical tracing of pilot balloons. Employing a searchlight is a geometrical method. The light fleck from the vertically operated searchlight is observed by a distant observer under a certain zenith angle. From the distance to the searchlight and the observation angle, the cloud base height can be computed from trigonometric relations. If modulated light is used the method may also be operated at daytime. Sometimes these instruments are also called ceilometers. Optical tracing of pilot balloons stops when they enter the cloud and are no longer visible for the tracing observer. The cloud base height is computed from the known ascent speed of the balloon.

9.2.2 Cloud cover

Cloud cover can automatedly be observed with a VIS/NIR Daylight Whole Sky Imager (Feister & Shields 2005). This instrument, which operates around 458, 664, and 878 nm at daytime only, detects the increased radiation from the sky in case of clouds with a fish-eye lens having a field of view of 180°. The amount of cloud cover may be derived from the surface by a statistical evaluation of the downward longwave radiation (LDR) from the clouds (Dürr & Philipona 2004). The longwave radiation emitted from clouds is characteristically larger than the one downwelling from clear skies. The method can be applied day and night and is as such superior to naked-eye observations. Determination of cloud cover data from satellites can be made globally but relies on complex evaluation algorithms (see e.g. Rossow & Schiffer (1991) and Jakob (2003)).

9.2.3 Cloud movement

An antiquated method to observe the cloud movement from the surface and by this also upper level winds (see Ch. 8.3.5) is the utilization of a nephoscope, i.e., a mirror with engravings with which the movement of clouds is followed (Braun 1881). For modern methods, the reader is referred to Chapter 8.3.5.

9.2.4 Water content

The integral cloud liquid water content (ILW) may be determined from satellites in a similar way as the integral water vapour content of the atmosphere (see Ch. 8.2.1) with multi-frequency microwave radiometers. The determination of ILW from the optical depth has an accuracy of between 10 and 20 % (Dabberdt et al. 2004). Martin et al. (2006b) give the following regression relations for observations at two frequencies (23.6 and 31.5 GHz):

$$ILW = -0.1431 - 1.4139\ \tau_{23.6\ GHz} + 5.6951\ \tau_{31.5GHz} \tag{9.6}$$

and at three frequencies (23.6, 31.5 und 151 GHz):

$$ILW = -0.0371 - 3.1300\ \tau_{23.6\ GHz} + 3.0342\ \tau_{31.5GHz} + 0.4406\ \tau_{151GHz}. \tag{9.7}$$

In Martin et al. (2006b), the accuracy for ILW is given with 0.014 kg/m^2.

9.3 Recommendations for remote sensing of liquid water and ice

As with remote sensing of atmospheric state variables (Ch. 8), surface-based observations are used for monitoring vertical distributions at certain locations or the horizontal distribution in the surrounding of the site, while satellite observations enable the survey of large parts of the Earth surface.

Weather RADAR and space-borne precipitation monitoring are well suited to capture the spatial structure of precipitation areas and their displacement speed. The quantative remote measurement of precipitation intensities is difficult because the drop size distribution is usually unknown. The measurement is best made with micro rain RADARs which allow for an estimation of this distribution.

Weather RADARs have become indispensable for short-term rain and thunderstorm predictions. Pre-alert times of 30 min up to a few hours are possible. Many countries have established weather RADAR networks which cover the whole territory. For routine nowcasting purposes, automated evaluation algorithms like the German program KONRAD (CONvection development in RADar products, Dotzek et al. 2007) have been developed. Basic products from these RADAR networks are broadcast in real time via the internet today.

The emission of stronger electromagnetic radiation may be harmful. RADAR instruments must be shielded by fences or by other means to prevent persons from coming too close to them. Simultaneously, such fences also serve to prevent the interference with other communication techniques operating at similar wavelengths. The operation of active electromagnetic remote sensing instruments often needs permits from the responsible authorities.

10 Remote sensing of trace substances

Remote sensing of trace substances may be based on three methods, either the detection of radiation absorption, emitted radiation, or by scattered radiation. A recent overview with an emphasis on satellite-based methods has been written by Burrows et al. (2007).

10.1 Trace gases

The wavelength-dependent absorption of radiation by trace gases may be utilized to measure the concentration of trace gases in the atmosphere. Depending on the physical properties of the gas molecules (i.e., the typical energy amounts these molecules can absorb by electronic transitions, excitations, and increased vibrational and/or rotational energy), different spectral ranges must be analysed to find these gases. For the detection of trace gases, analyses are made either at one single absorption line or over a wider range of spectra. For investigations concentrating on one absorption line, interference from other trace gases can become important and should be known. On the other hand, the spectral signature or fingerprint of a trace gas in a wider spectral range is rather unique. For spectral methods splitting of incoming radiation can be achieved either by a grating or prism spectrometer, or by in an interferometer.

In the UV range, we find, e.g., the absorption bands of ozone molecules (Hartley bands between 200 and 300 nm, a Huggins band at 340 nm, see Malicet et al. 1995). Oxygen (O_2) and nitrogen (N_2) molecules also have absorption bands in the UV range. O_2, H_2O, NO_2, ClO_2 and BrO have absorption bands in the visible range. All these absorption bands originate from electronic transitions.

In the infrared range, combined rotational and vibrational energy changes lead to absorption bands. The absorption frequency depends on the vibration frequency of the molecules. All atmospheric gases except diatomic gases have absorption bands in the infrared range and thus also contribute to the greenhouse effect which keeps the Earth warm.

In the far infrared (>40 μm), molecules with a permanent dipole moment have rotational absorption bands. These gases include H_2O, O_3, OH, CO, HCl, HF, HBr, N_2O, HCN, $HOCl$ HO_2, HNO_3 and ClO.

In the microwave range (> 1 mm), rotational bands of H_2O, O_3, H_2O_2, ClO and HNO_3 can be utilized for the detection of these trace gases.

10.1.1 Horizontal path-averaging methods

Path-averaging measurements in the infrared range are made with FTIR (Ch. 7.3.2) and in the UV and visible range with DOAS (Ch. 7.3.3). With FTIR (see also Tab. 24) CO, CO_2, CH_4, N_2O, NO, NO_2, SO_2, NH_3, SF_6, HCHO, BTX (benzine, toluene, xylene) as well as n-pentane, ethane, ethene, ethine, propane, propene, (iso-)butane, iso-butene, butadiene and hexane can be detected. To interpret infrared spectra of atmospheric measurements, a multi-component air pollution software (MAPS) was developed for retrieval of gas concentrations from radiation emission as well as absorption measurements (Schäfer et al. 1995). With DOAS (see also Tab. 25), BTX and other reactive gases such as O_3, NO_2, NO_3, (OH), HCHO, HONO and SO_2 can be detected (a comparison of different DOAS instruments is presented in Camy-Peyret et al. 1996).

10.1.2 Vertical column densities

Surface-based measurements of vertical column densities of ozone are made with a photometer (Ch. 7.4.2) that observes the sunlight at several pairs of neighbouring wavelengths (about 20 nm apart) in the Huggins band (310–350 nm). The pairs have been chosen so that at one wavelength ozone absorbs much more than at the other wavelength. The ozone column density can be derived from the absorption difference. The first measurements of the ozone column density were performed in 1921 (Fabry & Buisson 1921). Measurements at wavelength pairs allow for the elimination of influences from other absorbing atmospheric substances such as aerosols. Dobson spectrophotometers measuring wavelength pairs have been available since 1924 (Dobson & Harrison 1926). The instrument is called a Brewer spectrophotometer (Fig. 99, Brewer 1973) if a grate is used instead of a prism for the wavelength-dependent refraction of the light. The latter one was developed in the 1970s and has a better sensitivity than the former one. Both, a Dobson and a Brewer spectrophotometer are used as standard instruments (WMO 2006). Today, more than 100 Dobson and about 150 Brewer spectrophotometers are operated globally to monitor the

Fig. 99. Brewer spectrometer measuring ozone column densities at the Hohenpeißenberg observatory of the German Weather Service.

ozone layer. The detection principle is comparable to a DIAL (Ch. 7.2.5.3) but it is not an active method because the sun is the light source.

Since 1978, satellites have carried instruments for the monitoring of the ozone column density. The first one, TOMS (Total Ozone Mapping Spectrometer) on the satellite NIMBUS 7 had been working for 14 years and captured the reflected and scattered solar UV radiation in 6 channels from 312.5 to 380 nm by nadir soundings (McPeters & Labow 1996). The scanned swaths were so wide that nearly the whole Earth surface was covered every day. Presently, GOME (Global Ozone Monitoring Experiment) on board of the satellite ERS-2, which was launched in 1995, is active. This instrument operates in an enhanced spectral range from 240 to 790 nm so that other trace gases can also be monitored simultaneously. A comparison of data from Dobson and Brewer spectrophotometers and TOMS and GOME data has been made by Vanicek (2006). In 2004, the Ozone Monitoring Instrument (OMI) was launched aboard the EOS-Aura satellite. The OMI-TOMS algorithm uses just two wavelengths (317.5 nm and 331.2 nm under most conditions, and 331.2 nm and 360 nm for high ozone and high solar zenith angle conditions). The longer of the two wavelengths is used to derive the surface reflectivity. The shorter wavelength is strongly absorbed by ozone and is used to derive total ozone columns. The algorithm also calculates an aerosol index from the difference in surface reflectivity derived from the 331.2 nm and 360 nm measurements (Balis et al. 2007).

10.1.3 Sounding methods

Vertical sounding of trace gas profiles is possible with LIDARs. Because the intensity of the backscattered laser light depends on the absorption by gas molecules as well as on extinction and scattering due to aerosols, inversion procedures to isolate the fingerprint of the trace gases are necessary. A single measurement at just one wavelength is not sufficient if no other profile information is available. Therefore, usually measurements at least two wavelengths are performed for trace gas profile measurements (see e.g. Eisele & Trickl 2005).

Vertical ozone profiles can be obtained by surface-based observations with an ozone LIDAR (essentially a DIAL, see Ch. 7.2.5.3) such as the ALOMAR LIDAR (Arctic Lidar Observatory for Middle Atmosphere Research, see Skatteboe 1996) operated in Andøya (Norway). This DIAL emits light with a wavelength of 308 nm (there is strong absorption due to ozone at this wavelength) and with a wavelength of 353 nm (there is no absorption due to ozone). The latter wavelength is produced by a Raman cell. The instrument delivers ozone profiles in the upper troposphere and the stratosphere from 8 km up to about 50 km height with a vertical resolution of a few hundreds of metres (Hoppe et al. 1995). Another ozone LIDAR, a three frequency-DIAL operating at 277, 292 and 313 nm (Eisele et al. 1999) reaches heights from 150 m up to about 16 km with a vertical resolution of 22.5 m.

Vertical trace gas profiles may be retrieved from space by limb soundings (Fig. 100). For example, HALOE (HALogen Occultation Experiment) operating on the

10.2 Aerosols

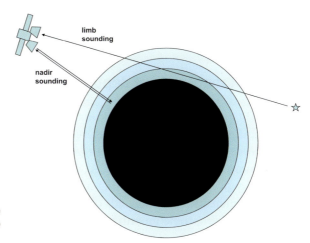

Fig. 100. Two modes of satellite sounding of the atmosphere: limb sounding and nadir sounding.

Upper Atmosphere Research Satellite (UARS) since 1991, captures vertical profiles of ozone, HCL, HF, NO, NO_2, CH_4, and aerosols. Likewise GOMOS (Global Ozone Monitoring by Occultation of Stars), operating on ENVISAT (ENVironmental SATellite) since 2002, measures profiles of ozone, NO_2, NO3, water vapour, and aerosols with several spectrometers and photometers at wavelengths between 250 and 952 nm by the observation of the occultation of bright stars in the horizon. On the same satellite, SCIAMACHY (SCanning Imaging Absorption spectroMeter for Atmospheric CHartographY) registers spectra between 240 and 2380 nm from subsequent limb and nadir soundings and thus delivers simultaneously at one location column densities and vertical profiles of ozone, NO, NO_2, BrO, OClO, ClO, water vapour, N_2O, CO, CO_2 and CH_4 as well as clouds and aerosols. A further device on this same satellite is MIPAS (Michelson Interferometer for Passive Atmospheric Sounding). MIPAS measures emissions from the atmosphere in limb soundings, too. From these soundings, vertical profiles of ozone, water vapour, HNO_3, CH_4, N_2O, and NO_2, as well as compounds such as HO_2NO_2, $ClONO_2$, N_2O_5, NO, and ClO, between 6 and 70 km height with a vertical resolution of 3 km are obtained by an inversion procedure (Hase & Fischer 2005).

10.2 Aerosols

The size of aerosol particles is in the range of a few nanometres to some micrometres. Because objects of this size interact most with radiation of comparable wavelengths, remote sensing of aerosols is mainly done by optical methods (cf. Fig. 62). An overview of different aerosol detection methods is given by Kahn et al. (2004). Aerosol characteristics may be derived from simultaneous measurements at different wavelengths, for example with a multi-spectral radiometer transmissometer (MSRT transmissometer, see, e.g., de Jong et al. 2007).

10.2.1 Aerosol optical depths (AOD)

The turbidity of the atmosphere is primarily a function of its aerosol mass concentration. Therefore, the quantitative measurement of the transparency of the atmosphere, or its inverse the aerosol optical depth (AOD), is a suitable mean for remote sensing of the atmospheric aerosol content. AOD is defined as that path length after which the intensity of radiation penetrating into the atmosphere has been reduced to 1/e of its original intensity. AOD depends on particle properties, size-dependent extinction efficiencies, hygroscopic growth and the vertical distribution of the particles.

10.2.1.1 Surface-based AOD measurements

The wavelength-dependent AOD, $\tau_A(\lambda)$ can be determined from sun photometer (Ch. 7.4.2) measurements of the total optical depth, $\tau(\lambda)$, if the effects of Rayleigh scattering by air molecules, $\tau_R(\lambda)$ and the absorption by ozone, $\tau_3(\lambda)$ and NO_2, $\tau_2(\lambda)$ are known:

$$\tau_A(\lambda) = \tau(\lambda) - \tau_R(\lambda) - \tau_3(\lambda) - \tau_2(\lambda). \tag{10.1}$$

AOD is measured with sun photometers in the AERONET network (AErosol RObotic NETwork, Holben et al. 1998) on a routine basis. In 1997, the network comprised about 60 globally distributed stations. From simultaneously performed measurements at several wavelengths, the Ångström exponent α can also be determined ($\tau_A(\lambda) = b\lambda^{-\alpha}$). The value of this exponent gives hints on the size distribution of the aerosol particles (Tanré et al. 2001). AOD measurements are possible only during daytime with a cloud-free sky. Even thin cirrus clouds can easily invalidate the measurements. But without additional independent information, it is difficult to detect such thin clouds.

10.2.1.2 Satellite-based AOD measurements

Radiometers on satellites receive shortwave radiation from the Earth which is composed of directly reflected and scattered sunlight. Scattering occurs at air molecules and at aerosol particles. The scattering properties of atmospheric aerosol particles depend on the wavelength of the scattered radiation and the scattering angle. Single scattering and, especially for larger optical thicknesses, multiple scattering contributes to the optical depth of the atmosphere (Conel 1990). For non-absorbing particles, the irradiation coming from multiple scattering is proportional to the single scattering contribution (King et al. 1999). An overview of aerosol measurements from satellites is presented by King et al. (1999) and Kaufman et al. (2002).

From measurements with single-channel radiometers, the increased intensity of the backscattered radiation or the reduced contrast between lighter and darker surface features due to the presence of particles in the atmosphere can be employed for the determination of AOD. For non-absorbing particles, the linear proportionality between AOD and particle backscatter is a good assumption. For absorbing particles, this relation is much more complex. With multi-spectral radiometers, column-integrated size distributions of aerosol particles may also be determined from the wavelength-dependent backscatter from particles with different sizes. Coarse-grained dust particles (e.g. Saharian dust) may be detected in the infrared range (10.5 to 12.5 µm) from an attenuation of the thermal emittance of the Earth's surface. This attenuation is due to absorption of thermal radiation by dust particles as well as to a cooling of the surface because of a dust-induced reduction of shortwave radiation reaching the surface. Also, the polarization of the backscattered radiation from particles yields information regarding their physical properties (Herman et al. 1997, Kaufman et al. 2002).

Satellite-based AOD measurements are available since 1981 from AVHRR (advanced very high resolution radiometer) data (Jacobowitz et al. 2003). Until 1994 (see Tab. 1 in King et al. 1999) AOD was based only on evaluations from the AVHRR sensor on several NOAA satellites and the TOMS (total ozone mapping spectrometer) sensor. Since 1994 evaluations from European satellites such as ERS-2 (European remote sensing satellite) and ENVISAT have also been available (King et al. 1999).

A more recent development is MODIS (Moderate Resolution Imaging Spectro-Radiometer) aboard the EOS (Earth observation system) satellites Terra and Aqua that makes observations at 36 different wavelengths. For the determination of AOD, observations at 0.47 (blue), 0.55 (green), and 0.66 (red) µm are utilized. An additional measurement at 2.1 µm determines the reflectivity of the surface because at this frequency, the dust-laden atmosphere is perfectly transparent. From the backscatter difference between the blue and the red channel, the particle size may be estimated. A detailed description of the MODIS algorithm may be found in Remer et al. (2005).

10.2.2 Sounding methods

10.2.2.1 Surface-based methods

Vertical profiles of atmospheric particle concentrations are retrievable from optical remote sensing with LIDARs or ceilometers (an example is given in Fig. 101). The analysis of these profiles requires a similar inversion procedure as the one for trace gases described in Chapter 10.1.3 in order to distinguish between the influences of the aerosols and trace gases on the intensity of the backscattered laser light. Ceilometers (Ch. 7.2.5.2) are especially suited for observations in the atmospheric boundary layer (Münkel et al. 2004). Because ceilometers are used only for soundings over

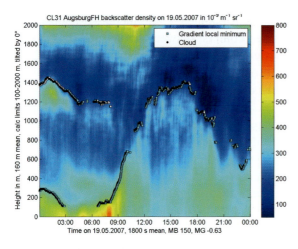

Fig. 101. Time-height cross-section of the optical backscatter intensity captured by a ceilometer. The backscatter intensity is mainly determined by the aerosol particle distribution. The dots designate the mixing height derived from an automated algorithm (see Fig. 96).

rather short distances, the influence of trace gases on the backscattered light intensity is neglected in the evaluation algorithms. With LIDARs (Ch. 7.2.5), the whole troposphere and even the lower stratosphere can be investigated (Reagan et al. 1989). The instruments which, e.g., are used in the European Aerosol LIDAR network (EARLINET), are described and compared in Matthais et al. (2004). The respective retrieval algorithms can be found in Böckmann et al. (2004). EARLINET is a network of 21 stations distributed over most of Europe using advanced quantitative laser remote sensing (multispectral backscatter LIDAR mostly combined with Raman LIDAR). Height attribution with these optical methods is made from the signal travel time. A Raman-LIDAR (Ch. 7.2.5.4) gives additional information on the particle properties such as size from extinction and depolarization measurements (see, e.g., Tesche et al. (2007), or for middle atmosphere studies, von Zahn et al. (2000)). The employment of a portable Raman polarization LIDAR for aerosol measurements at the southern edge of the Sahara desert is presented in Heese & Wiegner (2008).

10.2.2.2 Satellite-based methods

The best means for a space-borne detection of vertical aerosol profiles are satellite-based backscatter LIDARs operating at at least two wavelengths simultaneously (Kahn et al. 2004). An example of such instrumentation is the Geoscience Laser Altimeter System (GLAS) that is mounted on the NASA satellite ICESat (Ice, Cloud, and land Elevation satellite), which was launched in January 2003 (Zwally et a. 2002). It is based on a ND:YAG laser operating at 532 and 1064 nm. The vertical resolution of GLAS is 75 m, with an emission of 40 pulses per second. The diameter of the footprint is 70 m, the distance of two adjacent footprints is 170 m (Hoff et al. 2005). In April 2006, the French-American campaign "Cloud-Aerosol Lidar and Infrared Pathfinder Satellite Observations" (CALIPSO) was launched which delivers vertical aerosol profiles, too. This mission has, e.g., helped determine the heights at which atmospheric brown clouds occur. Atmospheric brown clouds are mostly the result of biomass burning and fossil fuel consumption. They consist of a mixture of light-absorbing and light-scattering aerosols and therefore contribute to atmospheric solar heating and surface cooling. They mostly appear between 1 and 3 km height above ground (Ramanathan et al. 2007).

10.3 Recommendations for remote sensing of trace substances

For all methods described in this chapter including horizontal path-averaging, vertical averaging and vertical sounding methods, the absorption characteristics of all possible trace gases must be known. A common source for such characteristics is, e.g., the HITRAN data base (http://cfa-www.harvard.edu/HITRAN/). The most recent 2008 edition of this database is described in Rothman et al. (2009).

Horizontally path-averaged trace substance concentrations are much better suited as input and evaluation data for numerical air quality and chemistry transport models than punctual in-situ data from instrumentation as described in Chapter 5, because such numerical models work with volume-averaged data and variables. When performing path-averaged measurements with optical methods, a compromise between the path length (a longer path length means more absorbing molecules and thus a higher detection sensitivity) and the received light intensity (a shorter path length means a higher received light intensity and thus an easier signal detection) must be found. Typically, horizontally path-averaged measurements are made over distances of several hundreds of metres.

Path-averaging methods are essentially bi-static methods which require the deployment of instruments and often an electrical supply for them at two separated sites. Even if both emitter and receiver are integrated into one instrument, a passive retro-reflector must still be positioned in some distance. The directional alignment of emitter and receiver must be done with great care and accuracy, because the receiver appears at a very small angle for the emitter. Even the daily temperature variation may cause the necessity of readjustments.

Vertical integrating methods are employed to get information required especially for radiation transfer calculations that determine the irradiation at the Earth surface. This includes the shortwave and longwave radiative balance at the surface, the determination of the anthropogenically induced greenhouse effect, as well as the assessment of the sunburn risk to human beings and plants by UV radiation.

Vertical profiles of trace substances must be retrieved for the assessment and understanding of radiative transfer as well as of chemical reactions taking place in the different levels of the atmosphere. Likewise, they are used as input and validation data for three-dimensional numerical chemistry transport models.

Technical rules for remote sensing may be found, e.g., in the VDI guidelines 4210 (DIAL), 4211 (FTIR), and 4212 (DOAS). FTIR methods are also treated in EN 15483:2008.

cies over 10 GHz cannot provide information from below the very first cm of the soil surface, especially when the soil is moist. Lower frequencies however are more sensitive to larger depths below the subsurface (Wagner et al. 1999, Prigent et al. 2005a).

The European Space Agency is currently planning the SMOS (soil moisture ocean salinity) mission (Kerr 2007). A novel passive instrument (MIRAS, Microwave Imaging Radiometer using Aperture Synthesis) has been developed that is capable of observing both soil moisture and ocean salinity by capturing images of emitted microwave radiation around the frequency of 1.4 GHz (L-band). This interferometric radiometer measures the brightness temperature of the Earth's surface at horizontal and vertical polarizations (Font et al. 2004).

11.1.4 Vegetation

Detection of vegetation parameters can be made by employing three different spectral properties of vegetation. At wavelengths shorter than 0.75 µm chlorophyll is absorbing strongly. In the near infrared range between 0.75 µm and 1.3 µm chlorophyll reflects quite well, while soil and water only weakly reflect in this spectral band. Above 1.3 µm, in three water absorption bands, radiation is absorbed by the water contained in the vegetation. The presence of vegetation and its state may be analysed from measurements in these three wavelength bands. The colour of the vegetation is detected in the visible range, the presence and state of the vegetation is analysed in the near infrared range, and the water content of the vegetation is analysed in the range between 1.3 and 2.6 µm. In contrast to naked soils chlorophyll's reflectivity suddenly increases at about 0.7 µm. This is utilized for the definition of the Normalized Difference Vegetation Index (NDVI, Fig. 102). The more active the

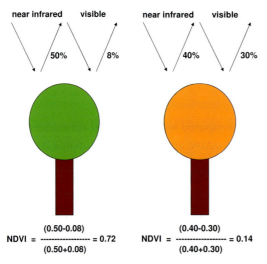

Fig. 102. Definition of the normalized difference vegetation index (NDVI). Fresh vegetation has a high NDVI, senescent vegetation has a low NDVI.

chlorophyll is, the stronger the increase of reflectivity. Denoting with r_1 the reflectivity between 0.55 μm and 0.75 μm and r_2 between 0.75 μm and 1.6 μm, the NDVI is defined as

$$NDVI = (r_2 - r_1) / (r_1 + r_2) \, . \tag{11.1}$$

The advantage of this definition is that by forming this ratio, the influence of surface geometry and the absorbing atmosphere are greatly reduced. When NDVI = 0, this signifies no vegetation at all, while fresh, young vegetation has a NDVI between 0.8 and 0.9. Senescent vegetation in autumn has a low NDVI at around 0.14. An overview of possible applications of NDVI in ecological studies is presented by Pettorelli et al. (2005).

11.1.5 Snow and ice

Different forms of snow and ice covered surfaces can be distinguished from backscattered radiation in the microwave range. A general overview of the remote sensing of snow and ice is given by Rees (2006). The emissivity of wet snow increases with increasing frequency from just under 0.9 to nearly 1.0 while that of dry snow decreases from 0.9 to about 0.75 in the frequency range between 10 and 100 GHz. The emissivity of new ice is similar to that of dry soil but for aged ice, the emissivity decreases with increasing frequency. Furthermore, the emissivity depends on the incident angle and the polarization. An overview of several possibilities to detect snow and ice from satellites has been written by König et al. (2001). Snow data products from the MODIS spectrometer (MODerate resolution Imaging Spectroradiometer) aboard the EOS Terra satellite can be obtained with a 500 m spatial resolution in the wavelength range between 0.4 and 14 μm (Hall et al. 2002).

11.1.6 Fires

Vulcanic eruptions and large fires (forest and bush fires, accidents in large industrial facilities) are easily detected from satellites in the middle infrared (Channel 3, see Tab. 21), because maximum radiation emission from fires is in this range (see e.g. Cuomo et al. 2001). Daytime Channel 3 satellite images look similar to images taken in the visible range, while nocturnal Channel 3 images have a stronger resemblance with infrared images. In both cases, hot spots can be detected in cloudfree conditions as bright spots. At daytime, additionally the smoke plumes are visible.

11.2 Properties of the ocean surface

The altitude and physical properties of the ocean surface, the salinity, and the sediment and algae content of the sea water are detectable from satellite observations. Introductions to ocean remote sensing and overviews can, e.g., be found in Johannessen et al. (2000) and Martin (2004).

11.2.1 Altitudes of the sea surface

With altimeters (Ch. 7.2.1.3) the altitude of the sea surface can be determined up to an accuracy of 2 to 3 cm. From altitude gradients of the sea surface, ocean circulation patterns may be derived (Wunsch & Stammer 1998). Likewise, the sea level rise due to Global Warming is detectable (Church et al. 2006).

11.2.2 Wave heights

Ocean wave heights may be obtained from space by altimetry or RADAR altimetry at 13.5 to 13.9 MHz (roughly 2 cm wavelength) (Bauer et al. 1992). The shape of the leading flank of the backscattered RADAR signal depends on wave height (Fig. 103). The RADAR pulse coming from above is first reflected at the wave crests and shortly afterwards from the troughs between the wave crests. Thus, the time duration

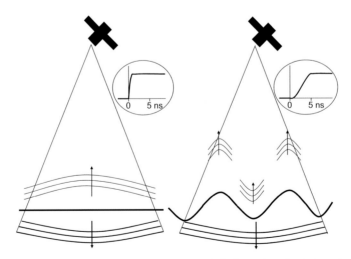

Fig. 103. Detection of wave height from satellite altimetry. Left: flat surface leads to a sharp leading edge of the return signal (displayed in the circle to the upper right). Right: wavy surface leads to a smoother leading edge of the return signal. The width of the leading edge is proportional to the wave height.

of the leading flank is longer the larger the trough-crest height difference. The accuracy of wave height measurements is a few centimetres; enough to detect even tsunamis in the open ocean (Smith et al. 2005). A method to derive horizontal fields of wave heights from satellite SAR data is presented in Schulz-Stellenfleth & Lehner (2004). From wave heights estimations of the near-surface wind speed (see also Ch. 8.3) is possible, although the relation between wave height and wind speed is not trivial (Emeis & Türk 2009).

11.2.3 Sea surface temperature

The sea surface temperature (SST) may be determined analogously to the land surface temperature (Ch. 11.1.2) from the brightness temperature, i.e. the radiant energy received in the near (thermal) infrared in cloudfree conditions (Fig. 94 middle). This requires a known emissivity of the sea surface. The accuracy of the method is about 0.5 K, the spatial resolution is about 1 km. An overview and a short history of infrared SST determination can be found in Barton (1995). It must be noted that satellite infrared sensors only observe the temperature of the skin of the ocean rather than the bulk SST traditionally measured from ships and buoys. The difference between the skin temperature and the true SST is usually 0.1 to 0.2 K but may amount up to 1 K (Schlüssel et al. 1990).

SST can also be determined from satellite microwave radiometry in all weather conditions except rain. Microwaves penetrate clouds with little attenuation, giving an uninterrupted view of the ocean surface. This is a distinct advantage over infrared measurements of SST, which are obstructed by clouds. Comparisons with ocean buoys show a root mean square difference of about 0.6 K (Wentz et al. 2000).

11.2.4 Salinity

The salinity of the sea water modifies the emissivity of the sea surface and thus the brightness temperature. If the sea surface temperature is known from additional measurements, the salinity may be determined from the received radiant energy. More reliable are probably microwave measurements. The microwave emissivity of sea water exhibits varying salinity dependencies at different frequencies (e.g. at 1.43 and 2.65 GHz, see Klein & Swift (1977)). Therefore, measurements at these two frequencies may be employed to determine the salinity (Lagerlöf et al. 1995). The abovementioned SMOS mission of ESA (see Ch. 11.1.3), which is scheduled to be launched in 2009, aims to performs salinity measurements with an accuracy of between 0.1 psu (200 km spatial resolution) and 0.5 psu (50 km spatial resolution) (Berger et al. 2002, Font et al. 2004).

11.2.5 Ocean currents

Large-scale ocean currents may be identified from altitude gradients of the sea surface if a geostrophic balance (i.e. a balance of pressure gradients forces and the Coriolis force) is assumed (see also Ch. 11.2.1). Direct measurements of the flow speed at the sea surface are not yet possible, but experiments are made to derive this speed from the Doppler shift in SAR soundings (Johannessen et al. 2008). It might though be possible to deduce flow patterns from sea surface temperature difference (Ch. 11.2.3) or differences in turbidity or algae content (Ch. 11.2.7), which may indicate the origin of the water masses (Johannessen et al. 2000).

11.2.6 Ice cover, size of ice floes

A general overview of the remote sensing of snow and ice is given by Rees (2006). In the visible and near infrared range, sea ice exhibits a considerably larger albedo than the open ocean water. In the visible range, it might sometimes be difficult to distinguish between sea ice and clouds. The use of AVHRR data for sea ice detection is discussed in Steffen et al. (1993).

The microwave emissivity of sea ice also differs significantly from that of open water. The brightness temperature of an ice surface in the microwave range is about 20 to 30 K higher than that of open water. An overview is presented in Carsey (1989). Measurements at two or more frequencies also allow for a distinction between new, first-year and multiyear ice. First-year ice has high emissivities for all frequencies while multiyear-ice emissivity decreases considerably with increasing frequency (Spreen et al. 2008).

A scatterometer (Ch. 7.2.1.4) is also suited to detect the extension of sea ice. The respective algorithm for the NASA scatterometer NSCAT, which operates in the Ku-band (see Tab. 23) and which was launched in 1996, is described in Remund & Long (1999).

11.2.7 Algae and suspended sediment concentrations

Algae and suspended sediment concentrations in the uppermost ocean layers may be analysed from the colour of the sea water in the visible range. Algorithms for the detection of phytoplankton are described in Sathyendranath et al. (1994) and O'Reilley et al. (1998). The detection of suspended sediment concentrations is discussed in Curran & Novo (1988). The difficulty in distinguishing between sediment and plankton concentrations is analysed in Tassan (1988).

12 Remote sensing of electrical phenomena

Thunderstorms are the most prominent electrical phenomenon in the atmosphere and are mainly recognizable from the occurrence of lightning and thunder. Remote sensing of atmospheric electrical phenomena usually involves the detection of the optical or electromagnetic signals originating from lightning (see Strangeways 2003 for a detailed treatment).

12.1 Spherics

Lightning discharges occur either between differently charged parts of a cloud (intra-cloud discharges, IC), or between clouds and the Earth's surface (cloud-to-ground discharges, CG). The electrical currents in a lightning stroke lead to the radiation of electromagnetic fields which are called "atmospherics" or even shorter "sferics" or "spherics" and which comprise a broad spectrum of frequencies. The detection of these sferics is a good means to trace the global thunderstorm activity (Ingmann et al. 1985). Measurements of sferics aim to determine two variables, the direction and the distance, from where the sferics originate (Wood & Inan 2002).

12.1.1 Directional analyses

Sferics direction is determined by analysis of the polarization of lightning-induced electromagnetic waves. With two loop-like or coil-like antennas oriented at a right angle to each other, the intensity of the sferics is measured. From the ratio of intensities from the two antennas, the direction of the thunderstorm can be detected with 180 degree ambiguity. Combining the measurements from several stations allows for a determination of the location of the thunderstorms. Sferics can be detected even over distances of several thousands of kilometres.

12.1.2 Distance analyses

The easiest determination of the distance to a thunderstorm is recording the time interval between the observation of the lightning flash and the arrival of the thunder. Due to the speed of sound in air, every three seconds corresponds to a distance of one kilometre (i.e., waiting fifteen seconds for the thunder means that the storm is still five kilometres away).

In a station network, the time-of-arrival method can be utilized. Here, the distance and location of the storm is computed inversely from the time differences in the reception of the signal of one and the same lightning strike at different stations. The accuracy is about 1 to 2 km inside a network and about 5 to 10 km outside. The European lightning network LINET (Betz et al. 2004) currently comprises about 80 stations recording sferics in the frequency range of 5 to 200 kHz. The network allows for a distinction between IC and CG discharges with a spatial resolution of about 200 m.

12.2 Optical lightning detection

A Lightning Imaging Sensor (LIS) is mounted on the satellite of the TRMM (Tropical Rainfall Measuring Mission) that records lightning flashes at a wavelength of 777.4 nm, an oxygen emission multiplet (a neutral oxygen line). LIS is a staring optical imager capturing an area of roughly 550 by 550 km, detecting rapid changes in the background scene. Spatial, spectral, and temporal filters determine whether these changes are associated with lightning. Horizontal resolution is 5 km, and a location is usually observed for 80 s during a satellite overpass. One flash in 80 s allows a minimum detectable flash rate of 0.7 min^{-1}. The detection efficiency is estimated to be about 85 % (Cecil et al. 2005), daytime thunderstorms are also detected. The satellite monitors more than half the Earth's surface, but can only observe part of the roughly 2000 thunderstorms which are active simultaneously at anyone time, and each of which on the average exhibits a lightning flash every 20 seconds (Strangeways 2003).

13 Outlook on new developments

Meteorological measurement techniques are a field of steady change and progress, both in in-situ techniques and in remote sensing techniques. This ongoing development is mainly driven by enhancements in electronics and microelectronics and in computer resources. One disadvantage of this development may be that a visual inspection of the instruments is no longer possible. The data quality has to be assessed a posteriori.

An example for the progress in in-situ techniques is the utilization of microelectronic techniques such as thin-film capacity sensors for moisture measurements or flexible diaphragms for pressure measurements. The ongoing miniaturization of the sensors helps to reduce the influence of the sensor on the atmospheric variable to be detected. Furthermore, miniaturization usually translates into less power consumption allowing for the remote deployment of instruments in places where no connection to an electrical grid is possible. Solar panels increasingly take over for power supply from batteries or diesel power engines in remote areas away from electrical grids. This even works for long-term monitoring installations. Computer and storage chip development enables the detection and storage of increasing amounts of data before download or transfer. Alternatively, real-time transmission of data via telecommunications is possible. In this way, maintenance intervals become longer and continuous operation becomes more feasible even in remote areas.

Even greater progress is currently being made in the field of remote sensing. Thus, the techniques and methods displayed in this book can only be a snapshot of the present state of the art. Further innovations can be expected in the next years. New laser technologies, miniaturization and reduced power consumption allows for additional spectral ranges in which satellite observations can be performed. Multispectral techniques may be utilized in ever growing fields of application. This further enhances the reliability of inversion procedures for the derivation of physical properties of the atmosphere and the gases, particles, and droplets within it, and of the Earth's surface.

This development can be illustrated by looking at the European Meteosat satellites. The first satellite of Meteosat Second Generation was launched in 2005 (Schmetz et al. 2002). But in parallel, development of the third generation of these satellites is under way (Stuhlmann et al. 2005). The third generation is scheduled to be in operation from 2015 onwards. Each satellite of the third generation is planned to have three imagers. One of these will operate with a high spatial resolution of (0.5 km in the subsatellite point) and a high repetition rate (every few minutes) for selected sections of the full image. The second imager will scan the full visible disc of the Earth with slightly reduced spatial and temporal (10 min) resolution but in several spectral channels. The third imager will be a lightning imaging system at

777.4 nm (see Ch. 12.2). Additionally, an infrared sounding system with 10 channels for vertical profiles of wind, temperature, humidity, and microphysical cloud parameters. The systems will be complemented by a sounder in the visible/UV range with 12 channels and 6 km spatial resolution in the subsatellite point for the analysis of atmospheric chemistry.

One of the classical atmospheric variables which still involves the largest difficulties even with in-situ measurements is precipitation. Problematic are the non-quantifiable influence of winds on rain gauge measurements and the extremely high spatial variability of precipitation in convective weather situations. Some hope comes with remote sensing. Optical precipitation sensors which detect drops falling through a light band do not influence the precipitation event and are not expected to be so susceptible to winds. A very recent research effort tries to investigate precipitation rates and areal distribution from the attenuation of radio waves in a wireless cellular telecommunication network (Messer 2007).

Another shift in measurement technology will probably come with modified applications for atmospheric data. An example is the growing exploitation of renewable energies which require special data which might not have been relevant so far. The operation of wind energy converters requires very accurate wind measurements, the operation of photovoltaic solar energy converters requires accurate shortwave radiation measurements, and the exploitation of water power needs accurate areal precipitation measurements. These accurate data are not only needed for an estimation of the amount of the energy output, but rather as input values for a short-term prediction of the temporal variation of the energy output. With the growing fraction of renewable energies in electricity generation, the perfect organisation of power grids connecting producers and consumers becomes increasingly important, because the production and the consumption of electrical energy must be balanced at any time.

The ultimate goal for many measurement tasks in atmospheric and environmental sciences and monitoring will probably be the symbiosis of in-situ methods, which can observe punctually but continuously, and remote sensing methods, which often (in the case of non-geostationary satellites) only give snapshots, but cover large areas or volumes. Also, numerical modelling will increasingly help to extend discontinuous measurements into continuous data coverage in space and time.

Literature

Abreu, V.J., J.E. Barnes, P.B. Hays, 1992: Observations of wind with an incoherent lidar detector. – Appl. Opt. **31**, 4509–4514.

Acharya, Y.B., A. Jayaraman, S. Ramachandran, B.H. Subbaraya, 1995: Compact light-emitting-diode sun photometer for atmospheric optical depth measurements. – Appl. Opt. **34**, 1209–1214.

Adrian, R.J., 1991: Particle-imaging techniques for experimental fluid mechanics. – Ann. Rev. Fluid Mech. **23**, 261–304.

Adriani, A., P. Massoli, G. Di Donfrancesco, F. Cairo, M.L. Moriconi, M. L., M. Snels, 2004: Climatology of polar stratospheric clouds based on lidar observations from 1993 to 2001 over McMurdo Station, Antarctica, – J. Geophys. Res. **109**, D24 211, doi:10.1029/2004JD004800.

Andreas, E.L., 1989: Two-wavelength method of measuring path-averaged turbulent surface heat fluxes. – J. Atmos. Oceanic Technol. **6**, 280–292.

Andreas, E.L., 2000: Obtaining Surface Momentum and Sensible Heat Fluxes from Crosswind Scintillometers. – J. Atmos. Oceanic Technol. **17**, 3–16.

Ansmann, A., M. Riebesell, C. Weitkamp, 1990: Measurement of atmospheric aerosol extinction profiles with a Raman lidar. – Opt. Lett. **15**, 746–748.

Archer, D., D. Stewart, 1995: The Installation and Use of a Snow Pillow to Monitor Snow Water Equivalent. – Water Env. J. **9**, 221–230.

Argawal, J.K., G.J. Sem, 1980: Continuous Flow, Single-Particle-Counting Condensation Nucleus Counter. – J. Aerosol Sci. **11**, 242–357.

Arnold, K.S, C.Y. She, 2003: Metal fluorescence lidar (light detection and ranging) and the middle atmosphere. – Contemp. Phys. **44**, 35–49.

Arnott, W.P., K. Hamasha, H. Moosmüller, P.J. Sheridan, J.A. Ogren, 2005: Towards Aerosol Light-Absorption Measurements with a 7-Wavelength Aethalometer: Evaluation with a Photoacoustic Instrument and 3-Wavelength Nephelometer. – Aerosol Sci. Technol. **39**, 17–29.

Arshinov, Yu.F., S.M. Bobrovnikov, V.E. Zuev, V.M. Mitev, 1983: Atmospheric temperature measurements using a pure rotational Raman lidar. – Appl. Opt. **22**, 2984–2990.

Asimakopoulos, D.N., C.G. Helmis, J. Michopoulos, 2004: Evaluation of SODAR methods for the determination of the atmospheric boundary layer mixing height. – Meteor. Atmos. Phys. **85**, 85–92.

Atlas, D. (Ed.), 1990: Radar in Meteorology: Battan Memorial and 40th Anniversary Radar Meteorology Conference. – Amer. Meteor. Soc., Boston. 806 pp.

Bacsik, Z., J. Mink, G. Keresztury, 2004: FTIR Spectroscopy of the Atmosphere. I. Principles and Methods. – Appl. Spectr. Rev. **39**, 295–363.

Balis, D., M. Kroon, M.E. Koukouli, E.J. Brinksma, G. Labow, J.P. Veefkind, R.D. McPeters, 2007: Validation of Ozone Monitoring Instrument total ozone column measurements using Brewer and Dobson spectrophotometer groundbased observations. – J. Geophys. Res. **112**, D24S46, doi:10.1029/2007JD008796.

Banakh, V.A., I.N. Smalikho, F. Köpp, C. Werner, 1995: Representativeness of wind measurements with a cw Doppler lidar in the atmospheric boundary layer. – Appl. Opt. **34**, 2055–2067.

Bange, J., R. Roth, 1999: Helicopter-borne flux measurements in the nocturnal boundary layer over land – a case study. – Bound.-Layer Meteorol. **92**, 295–325.

Baron, P.A., 1986: Calibration and Use of the Aerodynamic Particle Sizer (APS 3300). – Aerosol Sci. Technol. **5**, 55–67.

Barthelmie, R.J., L. Folkerts, F.T. Ormel, P. Sanderhoff, P.J. Eecen, O. Stobbe, N.M. Nielsen, 2003: Offshore Wind Turbine Wakes Measured by SODAR. – J. Atmos. Oceanic Technol. **20**, 466–477.

Barton, I.J., 1995: Satellite-derived sea surface temperatures: Current status. – J. Geophys. Res. **100**, 8777–8790.

Basist, A., N.C. Grody, T.C. Peterson, C.N. Williams, 1998: Using the Special Sensor Microwave/Imager to monitor land surface temperatures, wetness, and snow cover. – J. Appl. Meteorol. **37**, 888–911.

Bauer, E., S. Hasselmann, K. Hasselmann, H.C. Graber, 1992. Validation and assimilation of Seasat altimeter wave heights using the WAM wave model. – J. Geophys. Res. **97**, 12671–12683.

Baumbach, G., 1994: Luftreinhaltung. Entstehung, Ausbreitung und Wirkung von Luftverunreinigungen – Meßtechnik, Emissionsminderung und Vorschriften. Springer, Berlin Heidelberg etc., 461 pp.

Bean, B.R., E.J. Dutton, 1968: Radio meteorology. – Dover Publications, 435 pp.

Behrendt, A., J. Reichardt, 2000: Atmospheric temperature profiling in the presence of clouds with a pure rotational Raman lidar by use of an interference-filter-based polychromator. – Appl. Opt. **39**, 1372–1378.

Behrendt, A., T. Nakamura, M. Onishi, R. Baumgart, T. Tsuda, 2002: Combined Raman lidar for the measurement of atmospheric temperature, water vapor, particle extinction coefficient, and particle backscatter coefficient. – Appl. Opt. **41**, 7657–7666.

Bell, A.G., 1880: On the Production and Reproduction of Sound by Light. – Am. J. Sci. **20**, 305–324.

Bell, J.P., 1969: A new design principle for neutron soil moisture gages: the "Walligford" neutron probe. – Soil Sci. **108**, 160–164.

Bender, M., G. Dick, J. Wickert, T. Schmidt, S. Song, G. Gendt, M. Ge, M. Rothacher, 2008: Validation of GPS Slant Delays using Water Vapour Radiometers and Weather Models. – Meteorol. Z. **17**, 807–812.

Bendix, J., 2004: Geländeklimatologie. – Borntraeger, Berlin, Stuttgart, 282 pp.

Bennett, A.F., 2002: Inverse Modeling of the Ocean and Atmosphere. – Cambridge Univ. Press, Cambridge (UK), 256 pp.

Bennett, M., S. Christie, 2007: Doppler Lidar Measurements Using a Fibre Optic System. – Meteorol. Z. **16**, 469–477.

Bennett, M., H. Edner, R. Grönlund, M. Sjöholm, S. Svanberg, R. Ferrara, 2006: Joint application of Doppler Lidar and differential absorption lidar to estimate the atomic mercury flux from a chlor-alkali plant. – Atmos. Environ. **40**, 664–673.

Berger, M., A. Camps, J. Font, Y. Kerr, J. Miller, J. Johannessen, J. Boutin, M.R. Drinkwater, N. Skou, N. Floury, M. Rast, H. Rebhan, E. Attema, 2002: Measuring Ocean Salinity with ESA's SMOS Mission – Advancing the Science. – ESA Bulletin **111**, 113–121.

Bernabé, J.M., M.I. Carretero, E. Galán, 2005: Mineralogy and origin of atmospheric particles in the industrial area of Huelva (SW Spain). – Atmos. Environ. **39**, 6777–6789.

Bernhofer, C., C. Feigenwinter, T. Grünwald, R. Vogt, 2003: Spectral correction of water and carbon flux for EC measurements at the anchor station Tharandt. – TU Dresden, Tharandter Klimaprotokolle **8**, 1–13.

Betz, H.-D., K. Schmidt, W.P. Oettinger, M. Wirz, 2004: Lightning detection with 3D-discrimination of intracloud and cloud-to-ground discharges. – Geophys. Res. Lett. **31**, L11108. DOI: 10.1029/2004GL019821.

Bevis, M., S. Businger, S. Chiswell, T.A. Herring, R.A. Anthes, C. Rocken, R.H. Ware, 1994: GPS meteorology: Mapping zenith wet delays onto precipitable water. – J. Appl. Meteor. **33**, 379–386.

Beyrich, F., 1995: Mixing height estimation in the convective boundary layer using sodar data. – Bound.-Lay. Meteorol. **74**, 1–18.

Beyrich, F., 1997: Mixing height estimation from sodar data – a critical discussion. – Atmosph. Environ. **31**, 3941–3954.

Beyrich, F., U. Görsdorf, 1995: Composing the diurnal cycle of mixing height from simultaneous sodar and wind profiler measurements. – Bound.-Lay. Meteorol. **76**, 387–394.

Beyrich, F., H.A.R. De Bruin, W.M.L. Meijninger, J.W. Schipper, H. Lohse, 2002: Results from One-Year Continuous Operation of a Large Aperture Scintillometer over a Heterogeneous Land Surface. – Bound.-Lay. Meteorol. **105**, 85–97.

Beyrich, F., R.D. Kouznetsov, J.-P. Leps, A. Lüdi, W.M.L. Meijninger, U. Weisensee, 2005: Structure parameters for temperature and humidity from simultaneous eddy-covariance and scintillometer measurements. – Meteorol. Z. **14**, 641–649.

Birmili, W., F. Stratmann, A. Wiedensohler, 1999. Design of a DMA-based size spectrometer for a large particle size range and stable operation. – J. Aerosol Sci. **30**, 549–553.

Böckmann, C., U. Wandinger, A. Ansmann, J. Bösenberg, V. Amiridis, A. Boselli, A. Delaval, F. De Tomasi, M. Frioud, A. Hågård, M. Iarlori, L. Komguem, S. Kreipl, G. Larchevêque, V. Matthias, A. Papayannis, G. Pappalardo, F. Rocadenbosch, J. Schneider, V. Shcherbakov, and M. Wiegner, 2004: Aerosol lidar intercomparison in the frame of the EARLINET project. 2. Aerosol backscatter algorithms. – Appl. Opt. **43**, 977–989.

Bonner, C.S., M.C.B. Ashley, J.S. Lawrence, J.W.V. Storey, D.M. Luong-Van, S.G. Bradley, 2008: Snodar: a new instrument to measure the height of the boundary layer on the Antarctic plateau. Proc. of SPIE Vol. 7014, 70146I, 7 pp.

Bradley, S.G., 1999: Use of Coded Waveforms for SODAR Systems. – Meteorol. Atmos. Phys. **71**, 15–23.

Bradley, S.G., 2007: Atmospheric acoustic remote sensing. – CRC Press, 271 pp.

Bragg, W.H., W.L. Bragg, 1913: The Reflection of X-rays by Crystals. – Proc. Roy. Soc. Lond. A, **88**, 428–438.

Braun, C., 1881: Measuring the height of clouds. – Nature **23**, 458.

Breuer, L., H. Papen, K. Butterbach-Bahl, 2000: N_2O emission from tropical forest soils in Australia. – J. Geophys. Res. **105**, 26353–26367.

Brewer, A.W., 1973: A replacement for the Dobson spectrophotometer. – Pure Appl. Geophys. **106**, 919–927.

Brocks, F.V, S.J. Richardson, 2001: Meteorological Measurement Systems. – Oxford University Press, 290 pp.

Brown, S.S., H. Stark, S.J. Ciciora, R.J. McLaughlin, A.R. Ravishankara, 2002: Simultaneous in situ detection of atmospheric NO_3 and N_2O_5 via cavity ring-down spectroscopy. – Review of Scientific Instruments – September 2002 – Vol. **73**/9, 3291–3301.

Browning, K.A., R. Wexler, 1968: The Determination of Kinematic Properties of a Wind Field Using Doppler Radar. – J. Appl. Meteorol. **7**, 105–108.

Brüggemann, N., J.-P. Schnitzler, 2008: Comparison of Isoprene Emission, Intercellular Isoprene Concentration and Photosynthetic Performance in Water-Limited Oak (*Quercus pubescens* Willd. and *Quercus robur* L.) Saplings. – Plant Biology, **4**, 456–463.

Buck, A.L., 1976: The variable-path Lyman–alpha hygrometer and its operating characteristics. – Bull. Amer. Meteorol. Soc., **57**, 1113–1118.

Burrows, J., H. Fischer, K. Künzi, K. Pfeilsticker, U. Platt, A. Richter, M. Riese, G. Stiller, T. Wagner, 2007: Atmosphärische Spurenstoffe und ihre Sondierung. – Chem. Unserer Zeit **41**, 170–191.

Busch, N.E., L. Kristensen, 1976: Cup anemometer overspeeding. – J. Appl. Meteorol. **15**, 1328–1332.

Businger, J.A., S.P. Oncley, 1990: Flux measurement with conditional sampling. – J. Oceanic Atmosph. Technol. **7**, 349–352.

Butterbach-Bahl, K., R. Gasche, L. Breuer, H. Papen, 1997: Fluxes of NO and N2O from temperate forest soils: impact of forest type, N deposition and of liming on the NO and N2O emissions. – Nutrient Cycling in Agroecosystems **48**, 79–90.

Butterbach-Bahl, K., L. Breuer, R. Gasche, G. Willibald, H. Papen, 2002: Exchange of trace gases between soils and the atmosphere in Scots pine forest ecosystems of the northeastern German lowlands: 1. Fluxes of N2O, NO/NO2 and CH4 at forest sites with different N-deposition. – For. Ecol. Mangem. **167**, 123–134.

Byer, R.L., L.A. Shepp, 1979: Two-dimensional remote air pollution monitoring via tomography. – Optics Letters **4**, 75–77.

Campbell, G.S., B.D. Tanner, 1985: A krypton hygrometer for measurement of atmospheric water vapor concentration. – In: Moisture and humidity '85, International Symposium on Moisture and Humidity, Washington, D.C., April 15–18, 1985, Proceedings: Research Triangle Park, N.C., Instrument Society of America, pp. 609–614.

Camy-Peyret, C., B. Bergovist, B. Galle, M. Carleer, C. Clerbaux, R. Colin, C. Fayt, F. Goutail, M. Nunes-Pinharanda, J.P. Pommereau, M. Hausmann, U. Platt, I. Pundt, T. Rudolph, C. Hermans, P.C. Simon, A.C. Vandaele, J.M.C. Plane, N. Smith, 1996: Intercomparison of instruments for tropospheric measurements using differential optical absorption spectroscopy. – J. Atmos. Chem. **23**, 51–80.

Carsey, F., 1989: Review and status of remote sensing of sea ice. – IEEE J. Oceanic Eng. **14**, 127–138.

Carvalho, F., D. Henriques, 2000: Use of the Brewer spectrophotometer for aerosol optical depth measurements in the UV region. – Adv. Space. Res. **25**, 997–1006.

Cecil, D.J., E.J. Zipser, S.W. Nesbitt, 2005: Three Years of TRMM Precipitation Features. Part I: Radar, Radiometric, and Lightning Characteristics. – Mon. Wea. Rev. **133**, 543–566.

Chilson, P.B., T.-Y. Yu, R.G. Strauch, A. Muschinski, R.D. Palmer, 2003: Implementation and Validation of Range Imaging on a UHF Radar Wind Profiler. – J. Atmos. Oceanic Technol. **20**, 987–996.

Chow, T.L., 1994: Design and performance of a fully automated evaporation pan. – Agric. Forest Meteorol. **68**, 187–200.

Chu, X., G.C. Papen, 2005: Resonance fluorescence lidar for measurements of the middle and upper atmosphere. – In: Fujii, T., T. Fukuchi (Eds.): Laser Remote Sensing. CRC Press, Taylor & Francis, Boca Raton etc., pp. 179–432.

Chung, A., D.P. Chang, M.J. Kleeman, K.D. Perry, T.A. Cahill, D. Dutcher, E.M. McDougall, K. Stroud, 2001: Comparison of real-time instruments used to monitor airborne particulate matter. – J. Air Waste Manag Assoc. **51**, 109–120.

Church, J.A., N.J. White, J.R. Hunter, 2006: Sea-level rise at tropical Pacific and Indian Ocean islands. – Glob. Planet. Ch. **53**, 155–168.

Cimini, D., T.J. Hewison, L. Martin, J. Güldner, C. Gaffard, F.S. Marzano, 2006: Temperature and humidity profile retrievals from ground-based microwave radiometers during TUC. – Meteorol. Z. **15**, 45–56.

Clifford, S., G. Ochs, R. Lawrence, 1974: Saturation of optical scintillation by strong turbulence. – J. Opt. Soc. Amer. **64**, 148–154.

Cohn, S.A., W.M. Angevine, 2000: Boundary layer height and entrainment zone thickness measured by lidars and wind profiling radars. – J. Appl. Meteor. **39**, 1233–1247.

Colls, J., 2002: Air Pollution. 2nd edition. Clay's Library of Health and the Environment. Spon Press, London and New York. 560 pp.

Compton, D.A.C., J. Drab, H.S. Barr, 1990: Accurate infrared transmittance measurements on optical filters using an FT-IR spectrometer. – Appl. Opt. **29**, 2908–2912.

Conel, J.E., 1990: Determination of surface reflectance and estimates of atmospheric optical depth and single scattering albedo from Landsat Thematic Mapper data. – Intern. J. Rem. Sens. **11**, 783–828.

Cooley, C.W., J.W. Tuckey, 1965: An algorithm for the maschine calculation of complex Fourier series. – Math. Comput. **19**, 297–301.

Cooney, J., 1970: Remote measurements of atmospheric water-vapour profiles using the Raman component of laser backscatter. – J. Appl. Meteorol. **9**, 182–184.

Coulter, R.L., T.J. Martin, 1986: Results from a high power, high frequency sodar. – Atmos. Res. **20**, 157–269.

Crescenti, G.H., 1997: A look back on two decades of Doppler SODAR comparison studies. – Bull. Amer. Meteorol. Soc. **78**, 651–673.

Cuomo V., R. Lasaponara, V. Tramutoli, 2001: Evaluation of a new satellite-based method for forest fire detection. – Int. J. Remote Sensing **22**, 1799–1826.

Curran, P.J., E.M.M. Novo, 1988: The relationship between suspended sediment concentration and remotely sensed spectral radiance: a review. – J. Coastal Res. **4**, 351–368.

Dabberdt, W.F., G.L. Frederick, R.M. Hardesty, W.-C. Lee, K. Underwood, 2004: Advances in meteorological instrumentation for air quality and emergency response. – Meteorol. Atmos. Phys. **87**, 57–88.

Dash, P., F.-M. Göttsche, F.-S. Olesen, H. Fischer, 2002: Land surface temperature and emissivity estimation from passive sensor data: theory and practice-current trends. – Internat. J. Remote Sensing **23**, 2563–2594.

Davies, F., C.G. Collier, K.E. Bozier, G.N. Pearson, 2003: On the accuracy of retrieved wind information from Doppler lidar observations. – Quart. J. Roy. Meteor. Soc. **129**, 321–334.

de Bruin, H.A.R., 2002: Introduction: Renaissance of Scintillometry. – Bound.-Lay. Meteorol. **105**, 1–4.

de Hoffmann, E., V. Stroobant, 2007: Mass spectrometry: principles and applications. 3rd edition. Wiley-Interscience, 489 pp.

de Jong, A.N., A.M.J. van Eijk, P.J. Fritz, L.H. Cohen, M.M. Moerman, 2007: The use of multi-band transmission data collected at Scripps pier in November 2006 for the investigation of aerosol characteristics. – Proc. SPIE, Vol. 6708, 67080L.

Denmead, O.T., 2008: Approaches to measuring fluxes of methane and nitrous oxide between landscapes and the atmosphere. – Plant Soil **309**, 5–24.

Derr, V.E., C.G. Little, 1970: A Comparison of Remote Sensing of the Clear Atmosphere by Optical, Radio, and Acoustic Radar Techniques. – Appl. Opt. **9**, 1976–1992.

Desjardins, R.L., 1977: Description and evaluation of a sensible heat flux detector. – Bound.-Lay. Meteorol. **11**, 147–154.

Dobson, G.M.B., D.N. Harrison, 1926: Measurements of the amount of ozone in the Earth's atmosphere and its relation to other geophysical conditions. – Proc. R. Soc. London **A 110**, 660–693.

Dotzek, N., P. Lang, M. Hagen, T. Fehr, W. Hellmiss, 2007: Doppler radar observation, CG lightning activity and aerial survey of a multiple downburst in southern Germany on 23 March 2001. – Atmos Res. **83**, 519–533.

Douglas, C.A., L.L. Young, 1945: Development of a Transmissometer for Determining Visual Range. – NBS Technical Development Report No. 47.

Drescher, A.C., A.J. Gadgil, P.N. Price, W.W. Nazaroff, 1996: Novel approach for tomographic reconstruction of gas concentration distributions in air: use of smooth basis functions and simulated annealing. – Atmos. Environm. **30**, 929–940.

Dürr, B., R. Philipona, 2004: Automatic Cloud Amount Detection by Surface Longwave Downward Radiation Measurements. – J. Geophys. Res. **109** (D05201), doi:10.1029/2003JD004182.

Dürr, H.-P., 1988: Das Netz des Physikers. – Hanser, München, Wien, 490 pp.

Durst, F., A. Melling, J.H. Whitelaw, 1976: Principles and Practice of Laser Doppler Anemometry. – London, Academic Press.

DWD, 1973: Leitfäden für die Ausbildung im Deutschen Wetterdienst. Nr. 6, Instrumentenkunde. 2., vermehrte u. verbesserte Aufl. – Deutscher Wetterdienst, Offenbach am Main, 67 pp.

Egger, J., S. Bajrachaya, R. Heinrich, P. Kolb, S. Lämmlein, M. Mech, J. Reuder, W. Schäper, P. Shakya, J. Schween, H. Wendt, 2002: Diurnal Winds in the Himalayan Kali Gandaki Valley. Part III: Remotely Piloted Aircraft Soundings. – Mon. Wea. Rev. **130**, 2042–2058.

Eigenwillig, N., H. Fischer, 1982: Determination of midtropospheric wind vectors by tracking pure water vapor structure in METEOSAT water vapor image sequences. – Bull. Amer. Meteor. Soc. **63**, 44–57.

Eisele, H., T. Trickl, 2005: Improvements of the aerosol algorithm in ozone lidar data processing by use of evolutionary strategies. – Appl. Opt. **44**, 2638–2651.

Eisele, H., H.E. Scheel, R. Sladkovic, T. Trickl, 1999: High-Resolution Lidar Measurements of Stratosphere-Troposphere Exchange. – J. Atmos. Sci. **56**, 319–330.

Emeis, S., 2000: Who created Réaumur's thermometer scale? – Meteorol. Z. **9**, 185–187.

Emeis, S., M.J. Kerschgens, 1985: Sensitive Pressure Transducer to Deduce the Structure of Mesohighs. – Beitr. Phys. Atmosph. **58**, 407–411.

Emeis, S., K. Schäfer, 2006: Remote sensing methods to investigate boundary-layer structures relevant to air pollution in cities. – Bound-Lay. Meteorol. **121**, 377–385.

Emeis, S., M. Türk, 2004: Frequency distributions of the mixing height over an urban area from SODAR data. – Meteorol. Z. **13**, 361–367.

Emeis, S., M. Türk, 2009: Wind-driven wave heights in the German Bight. – Ocean Dyn. **59**, 463–475.

Emeis, S., Chr. Münkel, S. Vogt, W.J. Müller, K. Schäfer, 2004: Atmospheric boundary-layer structure from simultaneous SODAR, RASS, and ceilometer measurements. – Atmos. Environ. **38**, 273–286.

Emeis, S., M. Harris, R.M. Banta, 2007a: Boundary-layer anemometry by optical remote sensing for wind energy applications. – Meteorol. Z. **16**, 337–347.

Emeis, S., C. Jahn, C. Münkel, C. Münsterer, K. Schäfer, 2007b: Multiple atmospheric layering and mixing-layer height in the Inn valley observed by remote sensing. – Meteorol. Z. **16**, 415–424.

Emeis, S., K. Schäfer, C. Münkel, 2008: Surface-based remote sensing of the mixing-layer height – a review. – Meteorol. Z. **17**, 621–630.

Engelbart, D., 2005: Bodengebundene Fernerkundung am Meteorologischen Observatorium Lindenberg. – promet **31**, 134–147.

Engelbart, D.A.M., J. Bange, 2002: Determination of boundary-layer parameters using wind profiler/RASS and sodar/RASS in the frame of the LITFASS project. – Theor. Appl. Climatol. **73**, 53–65.

Engelbart, D., M. Kallistratova, R. Kouznetsov, 2007: Determination of the turbulent fluxes of heat and momentum in the ABL by ground-based remote-sensing techniques (a Review). – Meteorol. Z. **16**, 325–335.

English, S.J., 1999: Estimation of Temperature and Humidity Profile Information from Microwave Radiances over Different Surface Types. – J. Appl. Meteorol. **38**, 1526–1541.

Enting, I.G., 2002: Inverse Problems in Atmopsheric Constituent Transport. – Cambridge Univ. Press, Cambridge (UK), 408 pp.

Fabry, C., H. Buisson, 1921: Étude de l'extrémité ultraviolette du spectre solaire. – J. Physique **3**, 197–226.

Fassel, V.A., R.N. Kniseley, 1974: Inductively Coupled Plasma-Optical Emission Spectroscopy. – Anal. Chem. **46**, 1110–1116.

Feister, U., J. Shields, 2005: Cloud and radiance measurements with the VIS/NIR Daylight Whole Sky Imager at Lindenberg (Germany). – Meteorol. Z. **14**, 627–639.

Fellgett, P.B., 1951: Theory of Infra-red Sensitivites and Its Application to Investigations of Stellar Radiation in the Near Infra-red. – PhD Thesis Univ. Cambridge (UK).

Feltz, W.F., H.B. Howell, R.O. Knuteson, H.M. Woolf, H.E. Revercomb, 2003: Near Continuous Profiling of Temperature, Moisture, and Atmospheric Stability using the Atmospheric Emitted Radiance Interferometer (AERI). – J. Appl. Meteor. **42**, 584–597.

Fercher, A.F., W. Drexler, C.K. Hitzenberger, T. Lasser, 2003: Optical coherence tomography – principles and applications. – Rep. Progr. Phys. **66**, 239–303.

Fiocco, G., L.D. Smullin, 1963: Detection of Scattering Layers in the Upper Atmosphere /60-140 km/ by Optical RADAR. – Nature **199**, 1275–1276.

Flamant, C., J. Pelon, P.H. Flamant, P. Durand, 1997: Lidar determination of the entrainement zone thickness at the top of the unstable marin atmospheric boundary-layer. – Boundary-Layer Meteorol. **83**, 247–284.

Flesch, T.K., J.D. Wilson, E. Yee, 1995: Backward-time Lagrangian stochastic dispersion models and their application to estimate gaseous emissions. – J. Appl. Meteor. **34**, 1320–1322.

Foken, T., 2008: Micrometeorology. – Springer, Berlin, Heidelberg etc., 308 pp.

Foken, Th., R. Dlugi, G. Kramm, 1995: On the determination of dry deposition and emission of gaseous compounds at the biosphere-atmosphere interface. – Meteorol. Z., N. F. **4**, 91–118.

Font, J., G.S.E. Lagerloef, D.M. Le Vine, A Camps, O.-Z. Zanife, 2004: The determination of surface salinity with the European SMOS space mission. – IEEE Trans. Geosci. Rem. Sens. **42**, 2196–2205.

Forbes, J.M., Yu.I. Portnyagin, N.A. Makarov, S.E. Palo, E.G. Merzlyakov, X. Zhang, 1999: Dynamics of the lower thermosphere over South Pole from meteor radar wind measurements. – Earth Planets Space **51**, 611–620.

Foskett, L.W., N.B. Foster, W.R. Thickstun, R.C. Wood, 1953: Infrared absorption hygrometer. – Mon. Wea. Rev. **81**, 267–277.

Friedrich, K., M. Hagen, 2004: Wind Synthesis and Quality Control of Multiple-Doppler-Derived Horizontal Wind Fields. – J. Appl. Meteorol. **43**, 38–57.

Fruhstorfer, P., R. Niessner, 1994: Identification and Classification of Airborne Soot Particles Using an Automated SEM/EDX. – Mikrochim. Acta **113**, 239–250.

FZK, 2000: Wonach riecht's denn hier? – Pressemitteilung 12/2000 der Forschungszentrum Karlsruhe GmbH.

Gardner, C.S., 2004: Performance capabilities of middle-atmosphere temperature lidars: comparison of Na, Fe, K, Ca, Ca+, and Rayleigh systems. – Appl. Opt. **43**, 4941–4956.

Geiger, H., W. Müller, 1928: Elektronenzählrohr zur Messung schwächster Aktivitäten. – Naturwiss. **16**, 617–618.

Gens, R., J.L. van Genderen, 1996: SAR interferometry : issues, techniques, applications. – Internat. J. Rem. Sens. **17**, 1803–1836.

Gerthsen, C., H. Vogel, 1993: Physik. Ein Lehrbuch zum Gebrauch neben Vorlesungen. 17. verb. u. erw. Aufl. – Springer, Berlin etc., XXVI + 944 pp. and 8 Tab.

Gilman, G.W., H.B. Coxhead, F.H. Willis, 1946: Reflection of sound signals in the troposphere. – J. Acoust. Soc. Am. **18**, 274–283.

Grabmer, W., M. Graus, C. Lindinger, A. Wisthaler, B. Rappenglück, R. Steinbrecher, A. Hansel, 2004: Disjunct eddy covariance measurements of monoterpene fluxes from a Norway spruce forest using PTR-MS. – Intern. J. Mass Spectrom. **239**, 111–115.

Graus, M., A. Hansel, A. Wisthaler, C. Lindinger, R. Forkel, K. Hauff, M. Klauer, A. Pfichner, B. Rappenglück, D. Steigner, R. Steinbrecher, 2006: A relaxed-eddy-accumulation method for the measurement of isoprenoid canopy-fluxes using an online gas-chromatographic technique and PTR-MS simultaneously. – Atmos. Environ. **40**, Suppl. 1, 43–54.

Greenhut, G.K., J.K.S. Ching, R. Pearson Jr., T.P. Repoff, 1984: Transport of Ozone by Turbulence and Clouds in an Urban Boundary Layer. – J. Geophys. Res., **89**, 4757–4766.

Griffiths, P.R., 1983: Fourier transform infrared spectroscopy. – Science **222**, 297–302.

Grund, C.J., R.M. Banta, J.L. George, J.N. Howell, M.J. Post, R.A. Richter, A.M. Weickmann, 2001: High-resolution Doppler lidar for boundary layer and cloud research. – J. Atmos. Oceanic Technol. **18**, 376–393.

Gubler, H., 1981: An inexpensive remote snow depth gauge based on ultrasonic wave reflection from the snow surface. – J. Glaciol. **27**, 157–163.

Güldner, J., D. Spänkuch, 2001: Remote Sensing of the Thermodynamic State of the Atmospheric Boundary Layer by Ground-Based Microwave Radiometry. – J. Atmos. Oceanic Technol. **18**, 925–933.

Guenther, A.B., A.J. Hills, 1998: Eddy covariance measurement of isoprene fluxes. – J. Geophys. Res. **103**(D11), 13,145–13,152.

Güsten, H., G. Heinrich, R.W.H. Schmidt, U. Schurath, 1992: A novel ozone sensor for direct eddy flux measurements. – J. Atmos. Chem., **14**, 73–84.

Hairston, P.P., J. Ho, F.R. Quant, 1997: Design of an Instrument for Real-time Detection of Bio-aerosols Using Simultaneous Measurement of Particle Aerodynamic Size and Intrinsic Fluorescence. – J. Aerosol Sci. **28**, 471–482.

Hall, D.K., G.A. Riggs, V.V. Salomonson, N.E. DiGirolamo, K.J. Bayr, 2002: MODIS snow-cover products. – Rem. Sens. Env. **83**, 181–194.

Hansel, A., A. Jordan, C. Warneke, R. Holzinger, A. Wisthaler, W. Lindinger, 1999: Proton-transfer-reaction mass spectrometry (PTR-MS): on-line monitoring of volatile organic compounds at volume mixing ratios of a few pptv. – Plasma Sources Sci. Technol. **8**, 332–336.

Hanson, R.K., J.M. Seitzman, P.H. Paul, 1990: Planar laser-fluorescence imaging of combustion gases. – Appl. Phys. **B 50**, 441–454.

Hardesty, R.M., L.S. Darby, 2005: Ground-based and airborne lidar. – Encyclopedia of Hydrologic Sciences. M.G. Anderson (Ed.), Wiley, 697–712.

Harren, F.J.M., G. Cotti, J. Oomens, S. te Lintel Hekkert, 2000: Photoacoustic Spectroscopy in Trace Gas Monitoring. – In: R.A. Meyers (Ed.): Encyclopedia of Analytical Chemistry, John Wiley & Sons Ltd, Chichester, 2203–2226.

Harris, M., G.N. Pearson, J.M. Vaughan, D. Letalick, C. Karlsson, 1998: The role of laser coherence length in continuous-wave coherent laser radar. – J. Modern Optics **45**, 1567–1581.

Harris, M., G. Constant, C. Ward, 2001: Continuouswave bistatic laser Doppler wind sensor. – Appl. Opt. **40**, 1501–1506.

Harrison L., J. Michalsky, J. Berndt, 1994: Automated multifilter shadow-band radiometer: instrument for optical depth and radiation measurement. – Appl. Opt. **33**, 5118–5125.

Hartge, K.H., R. Horn, 2008: Die physikalische Untersuchung von Böden. 4th ed. – Schweizerbartsche Verlagsbuchhandlung, Stuttgart, 180 pp.

Hase, F., H. Fischer, 2005: Satellitengestützte Fernerkundung atmosphärischer Spurenstoffe. – Promet **31**, 38–43.

Hashmonay, R.A., M.G. Yost, C.-F. Wu, 1999: Computed tomography of air pollutants using radial scanning path-integrated optical remote sensing. – Atmos. Environm. **33**, 267–274.

Hasse, L., M. Großklaus, K. Uhlig, P. Timm, 1998: A ship rain gauge for use in high wind speeds. – J. Atmos. Oceanic Technol. **15**, 380–386.

Haus, R., K. Schäfer, W. Bautzer, J. Heland, H. Mosebach, H. Bittner, T. Eisenmann, 1994: Mobile Fourier-transform infrared spectroscopy monitoring of air pollution. – Appl. Optics, **33**, 5682–5689.

Hayden, K.L., K.G. Anlauf, R.M. Hoff, J.W. Strapp, J.W. Bottenheim, H.A. Wiebe, F.A. Froude, J.B. Martin, D.G. Steyn, I.G. McKendry, 1997: The Vertical Chemical and Meteorological Structure of the Boundary Layer in the Lower Fraser Valley during Pacific '93. – J. Atmos. Environ. **31**, 2089–2105.

Heese, B., M. Wiegner, 2008: Vertical aerosol profiles from Raman polarization lidar observations during the dry season AMMA field campaign. – J. Geophys. Res. **113**, D00C11, doi:10.1029/2007JD009487.

Heland, J., J. Kleffmann, R. Kurtenbach, P. Wiesen, 2001: A New Instrument to Measure Gaseous Nitrous Acid (HONO) in the Atmosphere. – Environ. Sci. Technol. **35**, 3207–3212.

Herman, M., J.L. Deuzé, C. Devaux, P. Goloub, F.M. Bréon, D. Tanré, 1997: Remote sensing of aerosols over land surfaces including polarization measurements and application to POLDER measurements. – J. Geophys. Res. **102**, 17039–17049.

Herrmann, R., C.T.J. Alkemade, 1963: Chemical analysis by flame photometry. – Translated by P.T. Gilbert, Jr. Interscience Publishers New York and London, XIV + 644 pp.

Hill, L.E., 1923: Spec. Rep. Ser. med. Res. Coun. (Lond.) No. 73.

Hill, R.J., 1997: Algorithms for Obtaining Atmospheric Surface-Layer Fluxes from Scintillation Measurements. – J. Atmos. Oceanic Technol. **14**, 456–467.

Hirsch, L., 1996: Spaced-Antenna-Drift Measurements of the Horizontal Wind Speed Using a FMCW-Radar-RASS. – Beitr. Phys. Atmos. **69**, 113–117.

Hoff, R.M., S.P. Palm, J.A. Engel-Cox, J. Spinhirne, 2005: GLAS long-range transport observation of the 2003 California forest fire plumes to the northeastern US. – Geophys. Res. Lett. **32**, L22S08, doi:10.1029/2005GL023723.

Högström, U., 1988: Non-dimensional wind and temperature profiles in the atmospheric surface layer: a re-evaluation. – Bound.-Lay. Meteor. **42**, 55–78.

Holben, B.N., T.F. Eck, I. Slutsker, D. Tanré, J.P. Buis, A. Setzer, E. Vermote, J.A. Reagan, Y.J. Kaufman, T. Nakajima, F. Lavenu, I. Jankowiak, A. Smirnov, 1998: AERONET – a federated instrument network for aerosol characterization. – Rem. Sens. Environ. **66**, 1–16.

Höpfner, M., B.P. Luo, P. Massoli, F. Cairo, R. Spang, M. Snels, G. Di Donfrancesco, G. Stiller, T. von Clarmann, H. Fischer, and U. Biermann, 2006: Spectroscopic evidence for

NAT, STS, and ice in MIPAS infrared limb emission measurements of polar stratospheric clouds. – Atmos. Chem. Phys. **6**, 1201–1219.

Hoppe, U.-P., G. Hansen, D. Opsvik, 1995: Differential absorption lidar measurements of stratospheric ozone at ALOMAR: First results. – Eur. Space Agency Spec. Publ., ESA SP-370, 335–344.

Horn, H.-G., 2005: Anzahlkonzentrations- und Größenverteilungsmessung von Nanopartikeln in der Außenluft. – VDI-Berichte **1885**, 47–61.

Hsieh, C.-I., G.G. Katul, 1997: Dissipation methods, Taylor's hypothesis, and stability correction functions in the atmospheric surface layer. – J. Geophys. Res. **102**, 16391–16405.

Humphreys, E., R.W. Clayton, 1988: Adaption of back projection tomography to seismic travel time problems. – J. Geophys. Res. **93**, 1073–1085.

Ingmann, P., J. Schaefer, H. Volland, M. Schmolder, A. Manes, 1985: Remote sensing of thunderstorm activity by means of VLF sferics. – Pure Appl. Geophys. **123**, 155–170.

Jacobowitz, H., L.L. Stowe, G. Ohring, A. Heidinger, K. Knapp, N.R. Nalli, 2003: The Advanced Very High Resolution Radiometer Pathfinder Atmosphere (PATMOS) Climate Dataset: A Resource for Climate Research. – Bull. Amer. Meteor. Soc. **84**, 785–793.

Jacobson, M.Z., 2002: Atmospheric Pollution. Cambridge Univ. Press, Cambridge (UK), 399 pp.

Jakob, C., 2003: An improved strategy for the evaluation of cloud parameterizations in GCMS. – Bull. Amer. Meteorol. Soc. **84**, 1387–1401.

Jakowski, N., 2005: Aufbau und Sondierung der Ionosphäre. Promet **31**, 35–37.

Jiménez, R., M. Taslakov, V. Simeonov, B. Calpini, F. Jeanneret, D. Hofstetter, M. Beck, J. Faist, H. van den Bergh, 2004: Ozone detection by differential absorption spectroscopy at ambient pressure with a 9.6 µm pulsed quantum-cascade laser. – Appl. Phys. **B78**, 249–256.

Johannessen, J.A., B. Chapron, F. Collard, V. Kudryavtsev, A. Mouche, D. Akimov, K.-F. Dagestad, 2008: Direct ocean surface velocity measurements from space: Improved quantitative interpretation of Envisat ASAR observations. – Geophys. Res. Lett. **35**, L22608, doi:10.1029/2008GL035709.

Johannessen, O.M., S. Sandven, A.D. Jenkins, D. Durand, L.H. Pettersson, H. Espedal, G. Evensen, T. Hamre, 2000: Satellite earth observation in operational oceanography. – Coastal Eng. **41**, 155–176.

Johansson, T.B., R.E. van Grieken, J.W. Nelson, J.W. Winchester, 1975: Elemental trace analysis of small samples by proton induced x-ray emission. – Anal. Chem. **47**, 855–860.

Joss, J., A. Waldvogel, 1967: Ein Spektrograph für Niederschlagstropfen mit automatischer Auswertung. – Pure Appl. Geophys. **68**, 240–246.

Junkermann, W., 2001: An Ultralight Aircraft as Platform for Research in the Lower Troposphere: System Performance and First Results from Radiation Transfer Studies in Stratiform Aerosol Layers and Broken Cloud Conditions. – J. Atmos. Oceanic Technol. **18**, 934–946.

Kadygrov, E.N., D.R. Pick, 1998: The potential performance of an angular scanning single channel microwave radiometer and some comparisons with in situ observations. – Meteor. Appl. **5**, 393–404.

Kahn, R.A., J.A. Ogren, T.P. Ackerman, J. Bösenberg, R.J. Charlson, D.J. Diner, B.N. Holben, R.T. Menzies, M.A. Miller, J.H. Seinfeld, 2004: Aerosol Data Sources and Their Roles within PARAGON. – Bull. Amer. Meteorol. Soc. **85**, 1511–1522.

Kaimal, J.C., J.E. Gaynor, 1991: Another look at sonic thermometry. – Bound.-Lay. Meteorol. **56**, 401–410.

Kalshoven Jr., J.E., C.L. Korb, G.K. Schwemmer, M. Dombrowski, 1981: Laser remote sensing of atmospheric temperature by observing resonant absorption of oxygen. – Appl. Opt. **20**, 1967–1971.

Kaufman, Y.J., D. Tanré, O. Boucher, 2002: A satellite view of aerosols in the climate system. – Nature **419**, 215–223.

Keckhut, P., M.L. Chanin, A. Hauchecorne, 1990: Stratosphere temperature measurement using Raman lidar. – Appl. Opt. **29**, 5182–5186.

Kerr, Y.H., 2007: Soil moisture from space: Where we are? – Hydrogeo. J. **15**, 117–120.

Keskinen, J., K. Pietarinen, M. Lehtimäki, 1992: Electrical low pressure impactor. – J. Aerosol Sci. **23**, 353–360.

Keuken, M.P., C.A.M. Schoonebeek, A. van Wensveen-Louter, J. Slanina, 1988: Simultaneous sampling of NH_3, HNO_3, HCl, SO_2 and H_2O_2 in ambient air by a wet annular denuder system. – Atmos. Environ., **22**, 2541–2548.

Kindler, D., A. Oldroyd, A. MacAskill, D. Finch, 2007: An 8 month test campaign of the QinetiQ ZephIR system: preliminary results. – Meteorol. Z. **16**, 463–473.

King, M.D., Y.J. Kaufman, D. Tanré, T. Nakajima, 1999: Remote Sensing of Tropospheric Aerosols from Space: Past, Present, and Future. – Bull. Amer. Meteor. Soc. **80**, 2229–2259.

Kinsey, J.L., 1977: Laser induced fluorescence. – Ann. Rev. Phys. Chem. **28**, 349–372.

Kleffmann, J., J. Heland, R. Kurtenbach, J.C. Lörzer, P. Wiesen, 2002: A new instrument (LOPAP) for the detection of nitrous acid (HONO). – Environ. Sci. Pollut. Res. **9**, 48–54.

Klein, L.A., C.T. Swift, 1977: An improved model for the dielectric constant of sea water at microwave frequencies. – IEEE Trans. Antennas and Propag., **AP-25**, 104–111.

Kleinschmidt, E., 1935: Handbuch der meteorologischen Instrumente und ihrer Auswertung. Springer, Berlin, XV+733 pp.

Kleissl, J., J. Gomez, S.-H. Hong, J.M.H. Hendrickx, T. Rahn, W.L. Defoor, 2008: Large Aperture Scintillometer Intercomparison Study. – Bound.-Lay. Meteorol. **128**, 133–150.

Knight, C.A., L.J. Miller, 1998: Early Radar Echoes from Small, Warm Cumulus: Bragg and Hydrometeor Scattering. – J. Atmos. Sci. **55**, 2974–2992.

Knuteson, R.O., F.A. Best, N.C. Ciganovich, R.G. Dedecker, T.P. Dirkx, S. Ellington, W.F. Feltz, R.K. Garcia, R.A. Herbsleb, H.B. Howell, H.E. Revercomb, W.L. Smith, J.F. Short, 2004a: Atmospheric Emitted Radiance Interferometer (AERI): Part I: Instrument Design. – J. Atmos. Oceanic Technol. **21**, 1763–1776.

Knuteson, R.O., F.A. Best, N.C. Ciganovich, R.G. Dedecker, T.P. Dirkx, S. Ellington, W.F. Feltz, R.K. Garcia, R.A. Herbsleb, H.B. Howell, H.E. Revercomb, W.L. Smith, J.F. Short, 2004b: Atmospheric Emitted Radiance Interferometer (AERI): Part II: Instrument Performance. – J. Atmos. Oceanic Technol. **21**, 1777–1789.

König, M., J. Winther, E. Isaksson, 2001: Measuring snow and glacier ice properties from satellite. – Rev. Geophys. **39**, 1–28.

Kohsiek, W., 1982: Measuring C_T^2, C_q^2 und C_{Tq} in the unstable surface layer, and relations to the vertical fluxes of heat and moisture. – Bound.-Lay. Meteorol. **24**, 89–107.

Kohsiek, W., W.M.L. Meijninger, H.A.R. De Bruin, F. Beyrich, 2006: Saturation of the Large Aperture Scintillometer. – Bound.-Lay. Meteorol. **121**, 111–126.

Kouznetsov, R., 2009: The multi-frequency sodar with high temporal resolution. – Meteorol. Z. **18**, 169–173.

Kramm, G., 1995: Zum Austausch von Ozon und reaktiven Stickstoffverbindungen zwischen Atmosphäre und Biosphäre. – IFU Schriftenreihe Bd. 34, Wissenschafts-Verlag Dr. Maraun, Frankfurt/M., 269 pp.

Kreyszig, E., 1972: Statistische Methoden und ihre Anwendungen. 3. Aufl. Vandenhoeck & Ruprecht, Göttingen.

Kristensen, L., 1993: The cup anemometer and other exciting instruments. – Doctor thesis at the Technical University of Denmark in Lyngby. Risø National Laboratory, Roskilde, Denmark. Risø-R-615 (EN), 83 pp.

Kursinski, E.R., G.A. Hajj, J.T. Schofield, R.P. Linfield, K.R. Hardy, 1997: Observing Earth's atmosphere with radio occultation measurements using the Global Positioning System. – J. Geophys. Res. **102**, 23429–23465.

Lagerlöf, G.S.E., C.T. Swift, D.M. Le Vine, 1995: Sea Surface Salinity: The Next Remote Sensing Challenge. – Oceanogr. **8**, 44–50.

Landis, S.M., R.K. Stevens, F. Schaedlich, E.M. Prestbo, 2002: Development and Characterization of an Annular Denuder Methodology for the Measurement of Divalent Inorganic Reactive Gaseous Mercury in Ambient Air. – Environ. Sci. Technol. **36**, 3000–3009.

Lawless, P.A., 1996: Particle Charging Bounds, Symmetry Relations, and an Analytic Charging Rate Model for the Continuum Regime. – J. Aerosol Sci. **27**, 191–215.

Lawrence, R.S., G.R. Ochs, S.F. Clifford, 1972: Use of scintillation to measure average wind across a light beam. – Appl. Opt. **11**, 239–243.

Lee, C., Y.H. Lee, S.H. Park, K.W. Lee, 2006: Design and evaluation of four-stage low-pressure cascade impactor using electrical measurement system. – Part. Sci. Technol. **24**, 329–351.

Lee, X., W. Massman, B. Law (Eds.), 2004: Handbook of Micrometeorology. – Kluwer Academic Publishers Dordrecht, 250 pp.

Lenschow, D.H., 1995: Micrometeorological techniques for measuring biosphere – atmosphere trace gas exchange. *In* Matsen, A., R.C. Harris (Eds.): Biogenic trace gases. Measuring emissions from soil and water. Blackwell Sci., Cambridge, 126–163.

Lenschow, D.H., B.B. Stankov, R. Pearson Jr., 1981: Estimating the ozone budget in the boundary layer by use of aircraft measurements of ozone eddy flux and mean concentration. – J. Geophys. Res., **86**, 7291–7297.

Lenschow, D.H., J. Mann, L. Kristensen, 1994: How long is long enough when measuring fluxes and other turbulence statistics. – J. Atmosph. Oceanis Technol. **11**, 661–673.

Leuning, R., 2004: Measurements of trace gas fluxes in the atmosphere using eddy covariance: WPL corrections revisited. – In: Lee, X., W. Massman, B. Law (Eds.), 2004: Handbook of Micrometeorology. – Kluwer Academic Publishers Dordrecht, 119–132.

Liebethal, C., T. Foken, 2006: On the use of two repeatedly heated sensors in the determination of physical soil parameters. – Meteorol. Z. **15**, 293–299.

Linacre, E., B. Geerts, 1997: Climates and Weather Explained. – Routledge, London, 432 pp.

Linné, H., B. Hennemuth, J. Bösenberg, K. Ertel, 2007: Water vapour flux profiles in the convective boundary layer. – Theor. Appl. Climatol. **87**, 201–211.

Liou, K.N., 2002: An Introduction to Atmospheric Radiation. 2nd edition. International Geophysics Series, Vol. 84. Academic Press Amsterdam etc. 583 pp.

Liu, W.T., 2002: Progress in Scatterometer Application. – J. Oceanogr. **58**, 121–136.

Löffler-Mang, M., 1998: A laser-optical device for measuring cloud and drizzle drop size distributions. – Meteorol. Z. **7**, 53–62.

Löffler-Mang, M., M. Kunz, W. Schmid, 1999: On the Performance of a Low-Cost K-Band Doppler Radar for Quantitative Rain Measurements. – J. Atmos. Oceanic Technol. **16**, 379–387.

Löhnert, U., C. Crewell, S. Simmer, 2004: An Integrated Approach toward Retrieving Physically Consistent Profiles of Temperature, Humidity, and Cloud Liquid Water. – J. Appl. Meteorol. **43**, 1295–1307.

Lovelock, J.E., 1974: The electron capture detector. – J. Chromatography **A 99**, 3–12.

Lübken, F.-J., M. Zecha, J. Höffner, J. Röttger, 2004: Temperatures, polar mesosphere summer echoes, and noctilucent clouds over Spitsbergen (78°N). – J. Geophys. Res. **109**, D11203, doi:10.1029/2003JD004247.

MacCready, P.B., 1966: Mean Wind Speed Measurements in Turbulence. – J. Appl. Meteorol. **5**, 219–225.

Malicet, J., D. Daumont, J. Charbonnier, C. Parisse, A. Chakir, J. Brion, 1995: Ozone UV spectroscopy. II. Absorption cross-sections and temperature dependence. – J. Atmos. Chem. **21**, 263–273.

Marenco, A., V. Thouret, P. Nedelec, H. Smit, M. Helten, D. Kley, F. Karcher, P. Simon, K. Law, J. Pyle, G. Poschmann, R. von Wrede, C. Hume, T. Cook, 1998: Measurement of ozone and water vapor by Airbus in-service aircraft: The MOZAIC airborne program. An overview. – J. Geophys. Res. **103**, 25,631–25,642.

Marjamäki, M., J. Keskinen, D.-R. Chen, D.Y.H. Pui, 2000: Performance evaluation of the Electrical Low-Pressure Impactor (ELPI). – J. Aerosol Sci. **31**, 249–261.

Marshall, J.S., S.M. Palmer, 1948: The distribution of raindrops with size. – J. Meteorol. **5**, 165–166.

Marshall, J.M., A.M. Peterson, A.,A. Barnes, 1972: Combined Radar-Acoustic Sounding System. – Appl. Opt. **11**, 108–112.

Martin, L., M. Schneebeli, C. Mätzler, 2006a: ASMUWARA, a ground-based radiometer system for tropospheric monitoring. – Meteorol. Z. **15**, 11–17.

Martin, L., M. Schneebeli, C. Mätzler, 2006b: Tropospheric water and temperature retrieval for ASMUWARA. – Meteorol. Z. **15**, 37–44.

Martin, S., 2004: An Introduction to Ocean remote Sensing. – Cambridge Univ. Press, 426 pp.

Massman, W., 2004: Concerning the measurement of atmospheric trace gas fluxes with open- and closed-path eddy covariance system: The WPL terms and spectral attenuation. – In: Lee, X., W. Massman, B. Law (Eds.), 2004: Handbook of Micrometeorology. – Kluwer Academic Publishers Dordrecht, 133–160.

Matthais, V., V. Freudenthaler, A. Amodeo, I. Balin, D. Balis, J Bösenberg, A. Chaikovsky, G. Chourdakis, A. Comeron, A. Delaval, F. De Tomasi, R. Eixmann, A. Hågård, L. Komguem, S. Kreipl, R. Matthey, V. Rizi, J.A. Rodrigues, U. Wandinger, X. Wang, 2004: Aerosol lidar intercomparison in the framework of the EARLINET project. 1. Instruments. – Appl. Opt. **43**, 961–976.

Mauder, M., C. Liebethal, M. Göckede, J.-P. Leps, F. Beyrich, T. Foken, 2006: Processing and quality control of flux data during LITFASS-2003. – Bound.-Lay. Meteorol. **121**, 67–88.

Mauder, M., T. Foken, C. Bernhofer, R. Clement, J. Elbers, W. Eugster, T. Grünwald, B. Heusinkveld, O. Kolle, 2008: Quality control of CarboEurope flux data – Part 2: Inter-comparison of eddy-covariance software. – Biogeosci. **5**, 451–462.

Maynard, A.D., 2007: Nanotechnology: The Next Big Thing, or Much Ado about Nothing? – Ann. Occup. Hyg. **51**, 1–12.

McArthur, L.J.B., D.H. Halliwell, O.J. Niebergall, N.T. O'Neill, J.R. Slusser, and C. Wehrli, 2003: Field comparison of network Sun photometers. – J. Geophys. Res. **108**, 4596, doi:10.1029/2002JD002964.

McMurray, P.H., M.F. Shepherd, J.S. Vickery, 2004: Particulate Matter Science for Policy Makers. A NARSTO Assessment. – Cambridge Univ. Press.

McPeters, R.D., G.J. Labow, 1996: An assessment of the accuracy of 14.5 years of Nimbus 7 TOMS version 7 ozone data by comparison with the Dobson network. – Geophys. Res. Lett. **23**, 3695–3698.

Meixner, F., 1994: Surface exchange of odd nitrogen oxides. – Nova Acta Leopoldina **NF 70**, 299–348.

Menut, L., C. Flamant, J. Pelon, P.H. Flamant, 1999 : Urban boundary-layer height determination from lidar measurements over the Paris area. – Appl. Optics **38**, 945–954.

Menzel, W.P., 2001: Cloud Tracking with Satellite Imagery: From the Pioneering Work of Ted Fujita to the Present. – Bull. Amer. Meteor. Soc. **82**, 33–47.

Menzel, W.P., F.C. Holt, T.J. Schmit, R.M. Aune, A.J. Schreiner, G.S. Wade, D.G. Gray, 1998: Application of GOES-8/9 Soundings to Weather Forecasting and Nowcasting. – Bull. Amer. Meteor. Soc. **79**, 2059–2078.

Messer, H., 2007: Rainfall monitoring using cellular networks. – IEEE Signal Proc. Mag. **24**, 142–144.

Meyers, J.F., 1995: Development of Doppler Global Velocimetry as a flow diagnostics tool. – Meas. Sci. Technol. **6**, 769–783.

Michelson, H.A., G.L. Manney, F.W. Irion, G.C. Toon, M.R. Gunson, C.P. Rinsland, R. Zander, E. Mahieu, M.J. Newchurch, P.N. Purcell, E.E. Remsberg, J.M. Russel, H.C. Pumphrey, J.W. Waters, R.M. Bevilacqua, K.K. Kelly, E.J. Hintsa, E.M. Weinstock, E.-W. Chiou, W.P. Chu, M.P. McCormick, C.R. Webster, 2002: ATMOS version 3 water vapor measurements: Comparisons with observations from two ER-2 Lyman-a hygrometers, MkIV, HALOE, SAGE II, MAS, and MLS. – J. Geophys. Res. **107** (D3), 10.1029/2001JD000587.

Middleton, W.E.K., 1969: The Invention of Meteorological Instruments. – John Hopkins Press, Baltimore, 362 pp.

Miklós, A., P. Hess, Z. Bosóki, 2001: Application of acoustic resonators in photoacoustic trace gas analysis and metrology. – Rev. Sci. Instrum. **72**, 1937–1955.

Möller, D., 2003: Luft. – De Gruyter, Berlin, New York, 750 pp.

Moncrieff, J.B., J.M. Massheder, H. de Bruin, J. Elbers, T. Friborg, B. Heusinkveld, P. Kabat, S. Scott, H. Soegaard, A. Verhoef, 1997: A system to measure surface fluxes of momentum, sensible heat, water vapour and carbon dioxide. – J. Hydrol. **188–189**, 589–611.

Münkel, C., S. Emeis, W.J. Müller, K. Schäfer, 2003: Observation of aerosol in the mixing layer by a ground-based lidar ceilometer. – In: Remote Sensing of Clouds and the Atmosphere VII, K. Schaefer, O. Lado-Bordowsky, A. Comeron, R.H. Picard (eds.), Proc. of SPIE, Bellingham, WA, USA, Vol. **4882**, 344–352.

Münkel, C., S. Emeis, W.J. Müller, K. Schäfer, 2004: Aerosol concentration measurements with a lidar ceilometer: results of a one year measurement campaign. – In: Remote Sensing of Clouds and the Atmosphere VIII, K. Schäfer, A. Comeron, M. Carleer, R.H. Picard (eds.), Proceedings of SPIE, Bellingham, WA, USA, Vol. **5235**, 486–496.

Münkel, C., N. Eresmaa, J. Räsänen, A. Karppinen, 2007: Retrieval of mixing height and dust concentration with lidar ceilometer. – Bound.-Lay. Meteorol. **124**, 117–128.

Munson, M.S.B., F.H. Field, 1966: Chemical Ionization Mass Spectrometry. I. General Introduction. – J. Amer. Chem. Soc. **88**, 2621–2630.

Muschinski, A., V. Lehmann, L. Justen, G. Teschke, 2005: Advanced Radar Wind Profiling. – Meteorol. Z. **14**, 609–625.

Navas, M.J., A. M. Jiménez, G. Galán, 1997: Air analysis: determination of nitrogen compounds by chemiluminescence. – Atmos. Environ. **31**, 3603–3608.

Neštor, V., 1996: Investigation of wind-induced error of precipitation measurements using a three-dimensional numerical simulation. – Zürich, Geograph. Inst., ETH. XIV, 117 pp. (Zürcher geographische Schriften 63).

Nimmermark, S., 2001: Use of electronic noses for detection of odour from animal production facilities: a review. – Water Sci. Technol. **44**, 33–41.

Nowak, D., D. Ruffieux, J.L. Agnew, L. Vuilleumier, 2008: Detection of Fog and Low Cloud Boundaries with Ground-Based Remote Sensing Systems. – J. Atmos. Oceanic Technol. **25**, 1357–1368.

Offiler, D., 1994: The Calibration of ERS-1 Satellite Scatterometer Winds. – J. Oceanic Atmos. Technol. **11**, 1002–1017.

O'Reilley, J.E., S. Maritorena, B.G. Mitchell, D.A. Siegel, K.L. Carder, S.A. Garver, M. Kahru, C. McClain, 1998: Ocean color chlorophyll algorithms for SeaWiFS. – J. Geophys. Res. **103**, 24937–24953.

Padley, F.B., 1969: The use of a flame-ionisation detector to detect components separated by thin-layer chromatography. – J Chromatogr. **39**, 37–46.

Parkinson, C.L., 2003: Aqua: An Earth-Observing Satellite Mission to Examine Water and Other Climate Variables. – IEEE Trans. Geosci. Rem. Sens. **41**, 173–183.

Parlange, M.B., G.G. Katul, 1992: Estimation of the diurnal variation of potential evaporation from a wet bare soil surface. – J. Hydrol. **132**, 71–89.

Pattey, E., R. L. Desjardins, P. Rochette, 1993: Accuracy of the Relaxed Eddy-Accumulation technique, evaluated using CO_2 flux measurements. – Bound.-Lay. Meteorol. **66**, 341–355.

Pedersen, T.F., 2003: Development of a Classification System for Cup Anemometers – CLASSCUP. – Risø Nat. Lab., Roskilde, Report Risø-R-1348(EN), 45 pp.

Peters, G., 1991: SODAR – ein akustisches Fernmeßverfahren für die untere Atmosphäre. – Promet **21**, 55-62.

Peters, G., B. Fischer, T. Andersson, 2002: Rain observations with a vertically-looking Micro Rain Radar (MRR). – Boreal Env. Res. **7**, 353–362.

Peters, G., B. Fischer, H. Münster, M. Clemens, A. Wagner, 2005: Profiles of Rain Drop Size Distributions as retrieved by Micro Rain Radars. – J. Appl. Met. **44**, 1930–1949.

Pettorelli, N., J.O. Vik, A. Mysterud, J.-M. Gaillard, C.J. Tucker, N.C. Stenseth, 2005: Using the satellite-derived NDVI to assess ecological responses to environmental change. – Trends Ecol. Evol. **20**, 503–510.

Philip, J.R., 1961: The theory of heat flux meters. – J. Geophys. Res. **66**, 571–579.

Phillips, P.D., H. Richner, J. Joss, A. Ohmura, 1980: ASOND-78: An intercomparison of Väisälä, VIZ and Swiss radiosondes. – Pure Appl. Geophys. **119**, 259–277.

Plaisance, H., S. Sauvage, P. Coddeville, R. Guillermo, 1997: A Ccomparison of Precipitation Sensors Used on the Wet-Only Collectors. – Env. Monit. Assessm. **51**, 657–671.

Plate, E.J., E. Zehe (Eds.), 2008: Hydrologie und Stoffdynamik kleiner Einzugsgebiete. – Schweizerbartsche Verlagsbuchhandlung, Stuttgart, 364 pp.

Platt, U., D. Perner, 1984: Ein Instrument zur spektroskopischen Spurenstoffmessung in der Atmosphäre. – Fresenius Z. Anal. Chem. **317**, 309–313.

Platt, U., J. Stutz, 2008: Differential Optical Absorption Spectroscopy: Principles and Applications. – Springer, Berlin/Heidelberg, 500 pp.

Poppitz, W., R. Heidenreich, 2005: Ermittlung der diffusen Partikelemission und Korngrößenverteilung eines Asphaltmischwerkes. – VDI-Berichte **1885**, 37–46.

Possanzini, M, A. Febo, A. Liberti, 1983: New design of a high-performance denuder for the sampling of atmospheric pollutants. – Atmos. Environ. **17**, 2605–2610.

Press, W.H., S.A Teukolsky, W.T. Vetterling, B.P. Flannery, 1992: Numerical Recipes in FORTRAN (2nd edition). – Cambridge Univ. Press, 963 pp.

Prigent, C., F. Aires, W.B. Rossow, A. Robock, 2005a: Sensitivity of satellite microwave and infrared observations to soil moisture at a global scale: Relationship of satellite observations to in situ soil moisture measurements. – J. Geophys. Res. **110**, D07110, doi:10.1029/2004JD005087.

Prigent, C., I. Tegen, F. Aires, B. Marticorena, M. Zribi, 2005b: Estimation of the aerodynamic roughness length in arid and semi-arid regions over the globe with the ERS scatterometer. – J. Geophys. Res. **110**, D09205, doi:10.1029/2004JD005370.

Prihodko, L., S.N. Goward, 1997: Estimation of air temperature from remotely sensed surface observations. – Rem. Sens. Environ. **60**, 335–346.

Pruitt, W.O., A. Angus, 1960: Large weighing lysimeter for measuring evapotranspiration. – Trans. ASAE **3**, 13–18.

Raabe, A., K. Arnold, A. Ziemann, 2001: Near surface spatially averaged air temperature and wind speed determined by acoustic travel time tomography. – Meteorol. Z. **10**, 61–70.

Ramanathan, V., M.V. Ramana1, G. Roberts, D. Kim, C. Corrigan, C. Chung, D. Winker, 2007: Warming trends in Asia amplified by brown cloud solar absorption. – Nature **448**, 575–579.

Ranweiler, L.E., J.L. Moyers, 1974: Atomic Absorption Procedures for Analysis of Metals in Atmospheric Particulate Matter. – Environ. Sci. Technol. **8**, 152–156.

Rapp, M., G.E. Thomas, G. Baumgarten, 2007: Spectral properties of mesospheric ice clouds: Evidence for nonspherical particles. – J. Geophys. Res., **112**, D03211, doi:10.1029/2006JD007322.

Reagan, J.A., M.P. McCormick, J.D. Spinhirne, 1989: Lidar sensing of aerosols and clouds in the troposphere and stratosphere. – Proc. IEEE **77**, 433–448.

Rees, G., 2006: Remote sensing of snow and ice. – CRC Press, 285 pp.

Reitebuch, O., 1999: SODAR-Signalverarbeitung von Einzelpulsen zur Bestimmung hochaufgelöster Windprofile. – Schriftenreihe des Fraunhofer-Instituts für Atmosphärische Umweltforschung, Shaker Verlag GmbH Aachen, vol. 62, 178 pp.

Reitebuch, O., C. Werner, I. Leike, P. Delville, P.H. Flamant, A. Cress, D. Engelbart, 2001: Experimental Validation of Wind Profiling Performed by the Airborne 10-μm Heterodyne Doppler Lidar WIND. – J. Atmos. Oceanic Technol. **18**, 1331–1344.

Remer, L.A., Y.J. Kaufman, S. Mattoo, J.V. Martins, C. Ichoku, R.C. Levy, and R.G. Kleidman, D. Tanré, D.A. Chu and R.-R. Li, T.F. Eck, E. Vermote, B.N. Holben, 2005: The MODIS Aerosol Algorithm, Products, and Validation. – J. Atmos Sci. **62**, 947–973.

Remund, Q.P., D.G. Long, 1999: Sea ice extent mapping using Ku band scatterometer data. – J. Geophys. Res. **104**, 11515–11527.

Reuder, J., P. Brisset, M. Jonassen, M. Müller, S. Mayer, 2009: The Small Unmanned Meteorological Observer SUMO: A new tool for atmospheric boundary layer research. – Meteorol. Z. **18**, 141–147.

Riddering, J.P., L.P. Queen, 2006: Estimating near-surface air temperature with NOAA AVHRR. – Can. J. Rem. Sens. **32**, 33–43.

Roedel, W., 1992: Physik unserer Umwelt – Die Atmosphäre. – Springer Verlag Heidelberg etc., 457 pp.

Rosenthal, I., T. Bercoviv, 1976: A chemical actinometer for measurement of U.V. radiation intensity in the atmosphere. – Atmos. Environ. **10**, 1139–1140.

Ross, J., M. Sulev, 2000: Sources of errors in measurements of PAR. – Agric. Forest Meteorol. **100**, 103–125.

Rossow, W.B., R.A. Schiffer, 1991: ISCCP cloud data products. – Bull. Amer. Meteorol. Soc. **72**, 2–20.

Rossow, W.B., R.A. Schiffer, 1999: Advances in understanding clouds from ISCCP. – Bull. Amer. Meteorol. Soc. **80**, 2261–2287.

Rothman, L.S., I.E. Gordon, A. Barbe, D.C. Benner, P.F. Bernath, M. Birk, V. Boudon, L.R. Brown, A. Campargue, J.-P. Champion, K. Chance, L.H. Coudert, V. Dana, V.M. Devi, S. Fally, J.-M. Flaud, R.R. Gamache, A. Goldman, D. Jacquemart, I. Kleiner, N. Lacome, W.J. Lafferty, J.-Y. Mandin, S.T. Massie, S.N. Mikhailenko, C.E. Miller, N. Moazzen-Ahmadi, O.V. Naumenko, A.V. Nikitin, J. Orphal, V.I. Perevalov, A. Perrin, A. Predoi-Cross, C.P. Rinsland, M. Rotger, M. Šimečková, M.A.H. Smith, K. Sung, S.A. Tashkun, J. Tennyson, R.A. Toth, A.C. Vandaele, J. Vander Auwer, 2009: The HITRAN 2008 molecular spectroscopic database. – J. Quant. Spectr. Rad. Transf. **110**, 533–572.

Rouffieux, D., J. Nash, P. Jeannet, J.L. Agnew, 2006: The COST 720 temperature, humidity and cloud profiling campaign TUC. – Meteorol. Z. **15**, 5–10.

Rummel, U., C. Ammann, A. Gut, F.X. Meixner, M.O. Andreae, 2002: Eddy covariance measurements of nitric oxide flux within an Amazonian rain forest. – J. Geophys. Res. **107**(D20), 8050, doi:10.1029/2001JD000520.

Saltzman, B.E., 1960: Modified Nitrogen Dioxide Reagent for Recording Air Analyzers. – Anal. Chem. **32**, 135–136.

Sandwell, D.T., W.H.F. Smith, 1997: Marine gravity anomaly from Geosat and ERS 1 satellite altimetry. – J. Geophys. Res. **102**, 10039–10054.

Sassen, K., T. Chen, 1995: The lidar dark band: An oddity of the radar bright band. – Geophys. Res. Lett. **22**, 3505–3508.

Sassen, K., J.R. Campbell, J. Zhu, P. Kollias, M. Shupe, C. Williams, 2005: Lidar and triple wavelength Doppler radar measurements of the melting layer: A revised model for dark and bright band phenomena. – J. Appl. Meteorol. **44**, 301–312.

Sathyendranath, S., F.E. Hoge, T. Platt, R.N. Swift, 1994: Detection of phytoplankton pigments from ocean color: improved algorithms. – Appl. Opt. **33**, 1081–1089.

Saussure, H.B. de, 1783: Essais sur l'hygrométrie. – Neuchatel.

Schäfer, K., R. Haus, J. Heland, A. Haak, 1995: Measurements of atmospheric trace gases by emission and absorption spectroscopy with FTIR. – Ber. Bunsen Gesell. **99**, 405–411.

Schlüssel, P., W.J. Emery, H. Grassl, and T. Mammen, 1990: On the Bulk-Skin Temperature Difference and Its Impact on Satellite Remote Sensing of Sea Surface Temperature. – J. Geophys. Res. **95**, 13341–13356.

Schmetz, J., P. Pili, S. Tjemkes, D. Just, J. Kerkmann, S. Rota, A. Ratier, 2002: An Introductio to Meteosat Second Generation (MSG). – Bull. Amer. Meteor. Soc. **83**, 977–992.

Schmid, H. P., 1994: Source areas for scalars and scalar fluxes. – Bound.-Lay. Meteorol. **67**, 293–318.

Schmid, H. P., 2002: Footprint modeling for vegetation atmosphere exchange studies: a review and perspective. – Agric. Forest Meteorol. **113**, 159–183.

Schulz-Stellenfleth, J., S. Lehner, 2004: Measurement of 2-D sea surface elevation fields using complex synthetic aperture radar data. – IEEE Trans. Geosci. Rem. Sens. **42**, 1149–1160.

Seibert, P., F. Beyrich, S.-E. Gryning, S. Joffre, A. Rasmussen, P. Tercier, 2000: Review and intercomparison of operational methods for the determination of the mixing height. – Atmosph. Environ. **34**, 1001–1027.

Senff, C., J. Bösenberg, G. Peters, T. Schaberl, 1996: Remote Sesing of Turbulent Ozone Fluxes and the Ozone Budget in the Convective Boundary Layer with DIAL and Radar-RASS: A Case Study. – Contrib. Atmos. Phys. **69**, 161–176.

Shao, Y., M.R. Raupach, 1992: The Overshoot and Equilibration of Saltation. – J. Geophys. Res. **97**, 20559–20564.

She, C.Y., J.R. Yu, H. Latifi, R.E. Bills, 1992: High-spectral-resolution fluorescence light detection and ranging for mesospheric sodium temperature measurements. – Appl. Opt. **31**, 2095–2106.

Shrivastava, G.P., 2008: Surface Meteorological Instruments and Measurement Practice. – Atlantic Publishers & Distributors (P) Ltd., New Dehli, 464 pp.

Sicard, M., C. Pérez, F. Rocadenbosch, J.M. Baldasano, D. García-Vizcaino1, 2006: Mixed-Layer Depth Determination in the Barcelona Coastal Area From Regular Lidar Measurements: Methods, Results and Limitations. – Bound.-Lay. Meteorol. **119**, 135–157.

Sigrist, M.W., 1994: Air Monitoring by Spectroscopic Techniques. – Wiley-IEEE, 531 pp.

Simmons, W. R.; Wescott, J. W.; Hall, F. F., Jr., 1971: Acoustic echo sounding as related to air pollution in urban environments. – NOAA TR ERL 216-WPL 17, Boulder, CO, 77 pp.

Singer, W., R. Latteck, P. Hoffmann, J. Bremer, 2005: Bodengebundene Radarmethoden zur Untersuchung der mittleren Atmosphäre. – promet **31**, 44–49.

Siple, P.A., C.F. Passel, 1945: Measurements of dry atmospheric cooling in subfreezing temperatures. – Proc. Amer. Phil. Soc. **89**, 177–199.

Skatteboe, R., 1996: ALOMAR: atmospheric science using lidars, radars and ground based instruments. – J. Atmos. Terr. Phys. **58**, 1823–1826.

Slanina, J., G.P. Wyers, 1994: Monitoring of atmospheric components by automatic denuder systems. – Fresenius' J. Anal. Chem. **350**, 467–473.

Smith, B.C., 1996: Fundamentals of Fourier transform infrared spectroscopy. – CRC Press, 202 pp.

Smith, E.K., S. Weintraub, 1953: The constants in the equation for atmopsheric refractive index at radio frequencies. – J. Res. Natl. Bur. Stand. **50**, 39–41.

Smith, W.H.F., R. Scharroo, V.V. Titov, D. Arcas, B.K. Arbic, 2005: Satellite Altimeters Measure Tsunami. – Oceanogr. **18** (2), 11–13.

Smith, W.L., W.F. Feltz, R.O. Knuteson, H.E. Revercomb, H.M. Woolf, H.B. Howell, 1999: The Retrieval of Planetary Boundary Layer Structure Using Ground-Based Infrared Spectral Radiance Measurements. – J. Atmos. Oceanic Technol. **16**, 323–333.

Solheim, F,, J.R. Godwin, 1998: Passive ground-based remote sensing of atmospheric temperature, water vapor, and cloud liquid water profiles by a frequency synthesized microwave radiometer. – Meteorol. Z., N.F. **7**, 370–376.

Solheim F., J.R. Goodwin, E.R. Westwater, Y. Han, S.J. Keihm, K. Marsh, R. Ware, 1998: Radiometric profiling of temperature, water vapor, and cloud liquid water using various inversion methods. – Radio Sci. **33**, 393–404.

Sonntag, D.; 1994: Advancements in the field of hygrometry. – Meteorol. Z., N.F. **3**, 51–66.

Spaan, W.P., G.D. van den Abeele, 1991: Wind borne particle measurements with acoustic sensors. – Soil Technol. **4**, 51–63.

Spänkuch, D., W. Döhler, J. Güldner, A. Keens, 1996: Ground-based passive atmospheic sounding by FTIR emission spectroscopy. First results with EISAR. – Beitr. Phys. Atmosph. **69**, 97–111.

Spiess, T., J. Bange, M. Buschmann, P. Vörsmann, 2007: First application of the meteorological Mini-UAV 'M²AV'. – Meteorol. Z. **16**, 159–169.

Spreen, G., L. Kaleschke, G. Heygster, 2008: Sea ice remote sensing using AMSR-E 89-GHz channels. – J. Geophys. Res. **113**, C02S03, DOI: 10.1029/2005JC003384.

Steffen, K., W. Abdalati, J. Stroeve, 1993: Climate sensitivity studies of the Greenland ice sheet using satellite AVHRR, SMMR, SSM/I and in situ data. – Meteorol. Atmos. Phys. **51**, 239–258.

Stephens, G.L. et al. 2002: The CloudSat Mission and the A-Train. – Bull. Amer. Meteorol. Soc. **83**, 1771–1790.

Strangeways, I., 2000: Measuring the Natural Environment. – Cambridge Univ. Press, 365 pp.

Strangeways, I., 2003: Measuring the Natural Environment. Second edition. – Cambridge Univ. Press, 534 pp.

Strangeways, I., 2007: Precipitation. Theory, Measurement and Distribution. – Cambridge Univ. Press, 290 pp.

Stuhlmann, R., A. Rodriguez, S. Tjemkes, J. Grandell, A. Arriaga, J.-L. Bézy, D. Aminou, P. Bensi, 2005: Plans for EUMETSAT's Third Generation Meteosat geostationary satellite programme. – Adv. Space Res. **36**, 975–981.

Stull, R.B., 1988: An Introduction to Boundary Layer Meteorology. – Kluwer Acad. Publ. Dordrecht, 666 pp.

Tanré, D., Y.J. Kaufman, B.N. Holben, B. Chatenet, A. Karnieli, F. Lavenu, L. Blarel, O. Dubovik, L.A. Remer, A. Smirnov, 2001: Climatology of dust aerosol size distribution and optical properties derived from remotely sensed data in the solar spectrum. – J. Geophys. Res. **106**, 18205–18217.

Tassan, S., 1988. The effect of dissolved yellow substance on the quantitative retrieval of chlorophyll and total suspended sediment concentrations from remote measurements of water colour. – Int. J. Remote Sens. **9**, 787–797.

Tatarskii, V.I., 1971: The effect of the turbulent atmosphere on wave propagation. – Kefer Press, Jerusalem, 472 pp.

Tesche, M., A. Ansmann, D. Müller, D. Althausen, R. Engelmann, M. Hu, Y.-H. Zhang, 2007: Particle backscatter, extinction, and lidar ratio profiling with Raman lidar in south and north China. – Appl. Opt. **46**, 6302–6308.

Tetzlaff, G., K. Arnold, A. Raabe, A. Ziemann, 2002: Observations of area averaged near-surface wind- and temperature-fields in real terrain using acoustic travel time tomography. – Meteorol. Z. **11**, 273–283.

Thiermann, V., H. Grassl, 1992: The measurement of turbulent surface-layer fluxes by use of bichromatic scintillation. – Bound.-Lay. Meteorol. **58**, 367–389.

Todd, L., D. Leith, 1990: Remote sensing and computed tomography in industrial hygiene. – Am. Ind. Hyg. Ass. J. **51**, 224–233.

Todd, L., N. Ramachandran, 1994: Evaluation of algorithms for tomographic reconstruction of chemical concentrations in indoor air. – Am. Ind. Hyg. Ass. J. **55**, 403–417.

Török, S.B., R.E. van Grieken, 1994: X-ray Spectrometry. – Anal. Chem. **66**, 186R–206R.

Trampert, J., J.-J. Leveque, 1990: Simultaneous iterative reconstruction techique: Physical interpretation based on the generalized least squares solution. – J. Geophys. Res. **95**, 12553–12559.

Trickl, T., H. Vogelmann, 2004: A powerful widely tunable single-mode laser system for lidar sounding of water vapor throughout the free troposphere. – In: Reviewed and revised papers presented at the 22nd International Laser Radar Conference (ILRC 2004), Matera, Italy, 12–16 July 2004, G. Pappalardo, A. Amodeo (Eds.), ESA SP-561, 175–178.

Troitsky, A.V., K.P. Gaikovich, V.D. Gromov, E.N. Kadygrov, A.S. Kosov, 1993: Thermal sounding of the atmospheric boundary layer in the oxygen band center at 60 GHz. – IEEE Trans. Geosci. Remote Sensing **31**, 116–120.

Tsoulfanidis, N., 1995: Measurement and detection of radiation. 2nd edition. – Taylor & Francis, 614 pp.

Turner, D.D., R.A. Ferrare, L.A.H. Brasseur, W.F. Feltz, T.P. Tooman, 2002: Automated Retrievals of Water Vapor and Aerosol Profiles from an Operational Raman Lidar. – J. Atmos. Oceanic Technol. **19**, 37–50.

Uppala, S.M. et al. 2005: The ERA-40 re-analysis. – Quart. J. Roy. Meteorol. Soc. **131**, 2961–3012.

Vanicek, K., 2006: Differences between ground Dobson, Brewer and satellite TOMS-8, GOME-WFDOAS total ozone observations at Hradec Kralove, Czech. – Atmos. Chem. Phys. **6**, 5163–5171.

Varma, R., H. Moosmüller, W.P. Arnott, 2003: Toward an ideal integrating Nephelometer. – Opt. Lett. **28**, 1007–1009.

Vietinghoff, H., 2000: Die Verdunstung freier Wasserflächen – Grundlagen, Einflussfaktoren und Methoden der Ermittlung. – UFO Atelier für Gestaltung und Verlag, Allensbach. 113 pp.

Viher, M., 2006: A study of the modified refraction indices over the Alpine and Sub-Alpine region. – Meteorol. Z. **15**, 625–630.

Vogel, H., 1993: Physik – Ein Lehrbuch zum Gebrauch neben Vorlesungen (Gerthsen Vogel). 17. verb. u. erw. Aufl. – Springer, Berlin etc., 944 pp.

von Zahn, U., G. von Cossart, J. Fiedler, K.H. Fricke, G. Nelke, G. Baumgarten, D. Rees, A. Hauchecorne, K. Adolfsen, 2000: The ALOMAR Rayleigh/Mie/Raman lidar: objectives, configuration, and performance. – Ann. Geophysicae **18**, 815–833.

Wagner, W., G. Lemoine, H. Rott, 1999: A method for estimating soil moisture from ERS scatterometer and soil data. – Remote Sensing of Environment **70**, 191–207.

Wakimoto, R.M., R. Shrivastava (Eds.), 2003: Radar and Atmospheric Science: A Collection of Essays in Honor of David Atlas. – Amer. Meteor. Soc., Boston. Meteor. Monogr. Ser. **30**, 270 pp.

Walton, S., 1996: Micrometeorological measurements and modelling of NO_x and O_3 exchange above and below European forest canopies. – PhD thesis University of Manchester, Inst. of Sci. and Technol., 209+xvii pp.

Webb, E. K., G. I. Pearman, R. Leuning, 1980: Correction of flux measurements for density effects due to heat and water vapour transfer. – Quart. J. Roy. Meteorol. Soc. **106**, 85–100.

Weill, A., C. Klapisz, F. Baudin, 1986: The CRPE miniSodar: Applications in micrometeorology and in physics of precipitations. – Atmos. Res. **20**, 317–335.

Weitkamp, C. (Ed.), 2005: Lidar. Range-Resolved Optical Remote Sensing of the Atmosphere. – Springer Science + Business Media Inc. New York, 455 pp.

Wentz, F.J., C. Gentemann, D. Smith, D. Chelton, 2000: Satellite Measurements of Sea Surface Temperature Through Clouds. – Science **288**, 847–850.

Werle, P., R. Kormann, 2001: Fast chemical sensor for eddy-correlation measurements of methane emissions from rice paddy fields. – Appl. Optics **40**, 846–858.

Werle, P., F. Slemr, M. Gehrtz, C. Bräuchle, 1989: Quantum-Limited FM-Spectroscopy with a Lead-Salt Diode Laser. – Appl. Phys. **B 49**, 99–108.

Werle, P., R. Mücke, F. Slemr, 1993: The Limits of Signal Averaging in Atmospheric Trace-Gas Monitoring by Tunable Diode-Laser Absorption Spectroscopy (TDLAS). – Appl. Phys. **B 57**, 131–139.

Wesely, M., 1976: The combined effect of temperature and humidity on the refractive index. – J. Appl. Meteorol. **15**, 43–49.

Wesely, M.L., D.H. Lenschow, O. T. Denmead, 1989: Flux measurement techniques. – In: Lenschow, D.H., B.B. Hicks (Eds.): Global Tropospheric Chemistry: Chemical fluxes in the global atmosphere, pp. 31–46. NCAR Boulder, Colorado.

Westwater, E.R., 1997: Remote Sensing of Tropospheric Temperature and Water Vapor by Integrated Observing Systems. – Bull. Amer. Meteorol. Soc. **78**, 1991–2006.

Westwater, E.R., Y. Han, V.G. Irisov, V. Leuskiy, E.N. Kadygrov, S.A. Viazankin, 1999: Remote sensing of boundary layer temperature profiles by a scanning 5-mm microwave radiometer and RASS: comparison experiments. – J. Atmos. Oceanic Technol. **16**, 805–818.

Westwater, E.R., S. Crewell, C. Mätzler, 2005a: Surface-based Microwave and Millimeter wave Radiometric Remote Sensing of the Troposphere: a Tutorial. – IEEE Geoscience and Remote Sensing Society Newsletter March 2005, 16–33.

Westwater, E.R., S. Crewell, C. Mätzler, D. Cimini, 2005b: Principles of Surface-based Microwave and Millimeter wave Radiometric Remote Sensing of the Troposphere. – Newsletter of SIEM, Società Italiana di ElettroMagnetismo, Vol.**1**, N3, pp. 50-90.

Wexler, R., D.M. Swingle, 1946: Radar storm detection. – Bull. Amer. Meteor. Soc. **28**, 159–167.

White, A.B., P.J. Neiman, F.M. Ralph, D.E. Kingsmill, P.O.G. Persson, 2003: Costal Orographic Rainfall Processes Observed by Radar during the California Land-Falling Jets Experiment. – J. Hydrometeor. **4**, 264–282.

Wickert, J., G. Gendt, 2006: Fernerkundung der Erdatmosphäre mit GPS. – promet **32**, 176–184.

Wickert, J., T. Schmidt, 2005: Fernerkundung der mittleren Atmosphäre mit GPS-Radiookkultation. – promet **31**, 50–52.

Wiedensohler, A., D. Orsini, D.S. Covert, D. Coffmann, W. Cantrell, M. Havlicek, F.J. Brechtel, L.M. Russell, R.J. Weber, J. Gras, J.G. Hudson, M. Litchy, 1997: Intercomparison Study of the Size-Dependent Counting Efficiency of 26 Condensation Particle Counters. – Aerosol Sci. Technol. **27**, 224–242.

Wilczak, J.M., E.E. Gossard, W.D. Neff, W.L. Eberhard, 1996: Ground-based remote sensing of the atmospheric boundary layer: 25 years of progress. – Bound.-Lay. Meteor. **78**, 321–349.

Wilson, D.K., D.W. Thomson, 1994: Acoustic tomography monitoring of the atmospheric surface layer. – J. Atmos. Ocean Tech. **11**, 751–769.

WMO, 1990: Guide to Meteorological Observation and Information Systems at Aerodromes. WMO Nr. 731. (cited from: Strangeways 2003).

WMO, 2006: Guide to Meteorological Instruments and Methods of Observation. Preliminary seventh edition. WMO No. 8, Genf, 569 pp. Available from: http://www.wmo.ch/pages/prog/www/IMOP/publications/CIMO-Guide/WMO-No_8.pdf.

WMO, 2008: Guide to Meteorological Instruments and Methods of Observation. Seventh edition. WMO No. 8, Genf, 681 pp. Available from: http://www.wmo.int/pages/prog/www/IMOP/publications/CIMO-Guide/CIMO_Guide-7th_Edition-2008.html.

Wolfe, G.M., J.A. Thornton, V.F. McNeill, D.A Jaffe, D. Reidmiller, D. Chand, J. Smith, P. Swartzendruber, F. Flocke, W. Zheng, 2007: Influence of trans-Pacific pollution transport on acyl peroxy nitrate abundances and speciation at Mount Bachelor Observatory during INTEX-B. – Atmos. Chem. Phys. **7**, 5309–5325.

Wolfe, G.M., J.A. Thornton, R.L.N. Yatavelli, M. McKay, A.H. Goldstein, B. LaFranchi, K.-E. Min, R.C. Cohen, 2009: Eddy covariance fluxes of acyl peroxy nitrates (PAN, PPN and MPAN) above a Ponderosa pine forest. – Atmos. Chem. Phys. **9**, 615–635.

Woo, S.Y., Y.J. Lee, I.M. Choi, J.W. Choi, 2004: New weight-loading device for calibration of precise barometers. – Metrologia **41**, 8–11.

Wood, T.G., U.S. Inan, 2002: Long-range tracking of thunderstorms using sferic measurements. – J. Geophys. Res. **107**, 4553, doi:10.1029/2001JD002008.

Woodman, R. F., A. Guillén, 1974: Radar observations of winds and turbulence in the stratosphere and mesosphere. – J. Atmos. Sci. **31**, 493–505.

Wu, C.-F., M.G. Yost, R.A. Hashmonay, 1999: Experimental evaluation of a radial beam geometry for mapping air pollutants using optical remote sensing and computed tomography. – Atmos. Environm. **33**, 4709–4716.

Wu, T.-Z., 1999: A piezoelectric biosensor as an olfactory receptor for odour detection: electronic nose. – Biosens. Bioelectr. **14**, 9–18.

Wulfmeyer, V., 1998: Ground-based differential absorption lidar for water-vapor and temperature profiling: Requirements, development, and specifications of a high-performance laser transmitter. – Appl. Opt. **37**, 3804-3824.

Wulfmeyer, V., 1999a: Investigation of Turbulent Processes in the Lower Troposphere with Water Vapor DIAL and Radar–RASS. – J. Atmos. Sci. **56**, 1055–1076.

Wulfmeyer, V., 1999b: Investigations of humidity skewness and variance profiles in the convective boundary layer and comparison of the latter with large eddy simulation results. – J. Atmos. Sci. **56**, 1077–1087.

Wunsch, C., D. Stammer, 1998: Satellite altimetry, the marine geoid, and the oceanic general circulation. – Ann. Rev. Earth Planetary Sci. **26**, 219–253.

Wysocki, G., M. McCurdy, S. So, D. Weidmann, C. Roller, R.F. Curl, F.K. Tittel, 2004: Pulsed quantum-cascade laser-based sensor for trace-gas detection of carbonyl sulfide. – Appl. Opt. **43**, 6040–6046.

Yanagisawa, S., N. Yamate, S. Mitsuzawa, M. Mori, 1966: Continuous Determination of Nitric Oxide and Nitrogen Dioxide in the Atmosphere. – Bull. Chem. Soc. Jp. **39**, 2173–2178.

Zudock, F., 1998: Katalytisch aktive CuO-Membranen zur Selektivitätssteuerung von Metalloxid-Gassensoren. – Wiss. Berichte Forschungszentrum Karlsruhe GmbH, FZKA 6220.

Zwally, H.J., B. Schutz, W. Abdalati, J. Abshire, C. Bentley, A. Brenner, J. Bufton, J. Dezio, D. Hancock, D. Harding, T. Herring, B. Minster, K. Quinn, S. Palm, J. Spinhirne, R. Thomas, 2002: ICESat's laser measurements of polar ice, atmosphere, ocean, and land. – J. Geodyn. **34**, 405–445.

Subject index

β-radiation absorption 85

A
absorption spectroscopy 125
accuracy 24, 31
acoustic received echo (ARE) method 179
actinometer 99, 100
actinometer, chemical 100
active, measurement 17
aerodynamic particle sizer 89, 91
aerosol 1, 29, 73, 144, 177, 193
aerosol analyser, electrical 90
characteristics, aerosol 193
aerosol detector, electrical 89, 92
aerosol particle 193
aerosol profile 196
aerovane 55, 62
aethalometer 86
air pressure 14
air quality 1, 73
aircraft 18, 20, 124, 133
albedo 101
albedometer 102, 103
alcohol 31
algae 204
algae content 202
altimeter 48, 133, 202
altimetry 202
anemometer 30, 55
anemometer, cup 55, 17, 27, 40, 54, 58, 59, 63, 174,
anemometer, hot wire 61, 63
anemometer, propeller 62, 63
anemometer, sonic 17, 37, 40, 113, 118, 119
anemometer, ultrasonic 36, 38, 61, 62, 63
aneroid 48
anti-Stokes scattering 148
artificial nose 95
atmometer 119, 121
atmospheric emitted radiance interferometer 160, 164
atmospheric windows 127, 199
atomic absorption spctroscopy 92, 93
atomic emission spectroscopy 93
automatic thermal evolution 93
availability 26

B
balloon 6, 18
barograph 50
barometer 14–17, 47, 48
aneroid barometer 6, 19, 48–50, 53
barometer, electronic 51
barometer, mercury 6, 14, 48, 49
barometer, water 49
Bayesian method 125
beam focussing 150, 175
Beaufort scale 56
Bellani atmometer 119
Bergerhoff method 86
Beta attenuation monitor 87
bimetal 30
bimetallic strip 33, 43
biogenic emission 33, 108
biosensors 95
black carbon 93
bolometer 38
Bowen-ratio method 107, 112
Bragg-RASS 141, 142, 143, 165, 166, 180
Bragg condition 127, 136, 141
Brewer spectrophotometer 102, 191
bulb 38
buoys 18, 69, 203

C
Campbell Stokes sunshine recorder 104
capacitive method 43, 46
capacitive sensor 47
capillary tube 11, 16, 31, 38
carbon dioxide 73, 114
cascade impactor 17, 89
cavity-ring-down-spectrometry 75, 77
ceilometer 145, 177, 187, 195
chemical composition of particle 92
chemical ionisation mass spectroscopy 75
chemical ionization mass spectrometer 114
chemical methods 81
chemiluminescence 75, 76, 82, 114
chill 104
chromatography 84
class A pan 121
Clausius-Clapeyron equation 41
clear air turbulence 176
clockwork 27
cloud 29, 130, 143, 167, 176, 187
cloud, amount of 187

cloud, atmospheric brown 196
cloud base 145, 187
cloud cover 188
cloud droplet 184
cloud liquid water 187
cloud liquid water content, integral 189
cloud movement 188
cloud top temperature 187
cloud water path 187
cloud wind 176
colorimeter 83
colorimetry 76, 83
computed tomography 160
condensation counter 89
condensation nucleus counter 91
condensation particle counter 91
conditional sampling 117
conductometry 76, 83
constant-level-balloon 20
constant altitude plane position indicator 185
counter tube 96
cuvette 107, 108

D

data acquisition 2
data logger 27
daylight whole sky imager 188
deadweight gauge 52
deadweight tester 52
denuder 74, 75, 81, 83
denuder, thermal desorption 82
deposit gauge 85, 86
deposition 110
deposition velocity 111
detection limit 25
detector 10
determination limit 25
dew 65
dewar flask 52
dewpoint 41, 42, 45
dewpoint determination 43
dewpoint mirror 42, 46
DIAL 17, 143, 146, 166, 180, 192
DIAL, water vapour 180
differential mobility analyser 90
differential mobility particle sizer 89
differential mobility sizer 90
differential mobility particle sizer, twin 90

differential optical absorption spectroscopy (DOAS) 155
digital instrument 17
direct measurement 7
direct observation 13
disdrometers 64, 67
dissipation method 107, 113
distance constant 58
DOAS 6, 17, 128, 157, 191
Dobson spectrophotometer 191
Doppler global velocimetry 152
Doppler LIDAR 180
Doppler RADAR 171, 129, 139, 132, 143, 149, 170, 173, 175
Doppler-RASS 141, 142, 143, 165
Doppler-SODAR 136, 137, 139
drift 25
droplet spectrum 184
drop size 64
drop size distribution 188
drop sonde 68
drop spectrograph 67, 68
dual-Doppler-RADAR 171
dynamic chamber 109

E

eddy accumulation method 107, 117, 118
eddy correlation method 113
eddy covariance 113
eddy covariance measurement 115
eddy covariance method 107, 112, 115, 118
eddy covariance method, disjunct 107, 118
electrical low pressure impactor 89
electrometer 92
electron capture detection 76
electron capture detector 80
electronic nose 95
electron microscopy 94
elemental carbon 93
emission flux 110, 122
emission spectroscopy 92
error 24
ethalometer 85
evaporation 29, 107, 118, 119
evaporation pan 119
evaporimeter 119, 121
evapotranspiration 119, 120
eye safety 144, 145, 182

Index

F

Fabry-Perot interferometer 149
fast Fourier infrared (FTIR) absorption spectroscopy 153, 155
fast mobility particle sizer 90
field capacity 71
filter 75, 94
filter paper 17, 74, 86
filter tape 94
fires 201
five-hole-sonde 63
five-hole probe 60, 63
fixed echo 135, 140
flame ionisation detection 76, 78, 80
flame photometry 76, 78
flexible diaphragm 207
fluorescence 76, 78, 127, 143
fluorescence aerodynamical particle sizer 91
flux 3, 9, 38, 77, 98, 106, 115, 119, 153
flux, mean 9
flux variance method 107, 112
fog 65
footprint 115, 122
forward model 14
Frankenberger psychrometer 45
frequency-domain interferometry 174
frigorimeter 99, 104
FTIR 128, 153, 157, 164, 191

G

gas analyser, infrared 114
gas analyser, optical 114
gas analyzer, residual 75
gas chromatograph 79, 80
gas chromatography 75, 76, 80, 118
gas chromatography mass spectrometry 75
Geiger counter 96
GPS radio occultation method 166
GPS receiver 20
GPS signal 168
gradient method 107, 111
gravimetric method 70, 72
ground clutter 130
gust 174
gustiness 54

H

hail 64, 65, 66, 186
helipod 18
high pressure/liquid chromatography 75
homogeneity 23
hot wire 55
humidity 40, 167, 171
humidity, absolute 41
humidity, relative 41, 42, 46
humidity, saturation 41
humidity, specific 30, 41
humidity profile 167, 168
humidity sensor, capacitive 46
hydrometeor 129
hyetometer 64
hygrograph 43
hygrometer 42, 43, 75
hygrometer, capacitive 46
hygrometer, hair 6, 17, 43, 44, 46
hygrometer, infrared 114
hygrometer, infrared absorption 44
hygrometer, krypton 44
hygrometer, lithium chloride 19, 44
hygrometer, Lyman-alpha- 44
hygrometer, optical 42, 44, 46
hygrometer, UV 44, 114
hypsometer 17, 48, 50
hysteresis 25

I

ice 40, 64, 201, 204
ice content 188
ideal gas 3, 41
ill-posed problem 16
imaging 8, 127
imaging method 157
impactor 75
indirect method 7, 12
inferential method 107, 110, 111
interferometer, infrared 167
interferometry, infrared 166
in-situ measurement 7, 29
in-situ method 74
instantaneous measurement 8
instrument, active 16
instrument, analogue 17
integral water vapour content 167
integrating measurement 8
interference 26

interferometer 154
interferometric procedure 132
internal energy 29
inverse method 107, 163
inverse model 14
inverse modelling 13, 122
inversion 12, 15
ionospheric electron density 181
irradiance, diffuse shortwave 100
irradiance, longwave 103
isoprene 114

K
katathermometer 99, 104

L
laser ablation mass spectrometry 75
laser Doppler anemometry 151
laser Doppler velocimetry 151
laser technology 207
laser-induced fluorescence 75
LIDAR 106, 126, 128, 143, 170, 173, 177, 186, 187, 192, 195, 196
LIDAR, airborne wind 175
LIDAR, backscatter 143–145, 196
LIDAR, differential absorption (DIAL) 143, 146
LIDAR, ozone 192
LIDAR, Raman 143, 147, 148, 166, 168, 196
LIDAR, resonance backscatter 148
LIDAR, resonance fluorescence 143, 148, 166, 177
LIDAR, water vapour 168
LIDAR, wind 172
lightning 205
lightning imaging sensor 206
limb sounding 21, 192
limiting condition of operation 26
liquid chromatography 76, 80
liquid thread 11, 13, 16
liquid water content 188
liquid-in-glass 30
Livingstone atmometer 119
long path absorption photometry 76, 83
low-pressure impactor 89
Lyman-α-resonance-fluorescence 75
lysimeter 119–121

M
Magnus formula 42
Malvern particle sizer 68
manometer 47, 60
Marshall-Palmer distribution 184
mass balance method 107, 110
mass spectrometer 79, 80
mass spectrometry 75, 76, 79
mass spectroscopy 84
mast 18, 177
measurement platform 18
mercury 16, 31, 48
mercury pile 14, 15
mercury thread 14
meter 43, 48
meter stick 7, 11, 68
methods, active 186
Michelson Interferometer 153, 187, 193
microbarometer 48, 52
microelectronic 4, 207
micrometeorological mast 115
micrometeorological method 10, 121
micrometeorological technique 106
micro rain RADAR 132, 186, 189
microscope 94
microwave method 163
microwave range 127
microwave region 159
microwave scintillometer 153
Mie scattering 127, 143, 159
miniaturization 207
miniSODAR 139
mixing-layer height 177
mixing ratio 41
moisture 29
moisture, absolute 40
moisture, saturation 41
moisture, specific 36
moisture flux 121, 180
monoterpene 118
mountain observatory 6
MOZAIC 20
multi-spectral radiometer transmissometer (MSRT transmissometer) 194
multiband transmissometer 106
multifilter rotating shadowband radiometer 102
multispectral technique 207

Index

N

nadir sounding 21
negative temperature coefficient 34
nephelometer 85, 86, 106
nephoscope 188
net pyrgeometer 103
net radiometer 103
neutron probe 70
nitrate 114
nitric oxide 82, 114
nitrogen dioxide 159
noctilucent cloud (NLC) 187
normalized difference vegetation index 163, 200

O

observation 2
observational network 6
observer 26
occultation 193
occultation measurement 22
ocean altitude 198
ocean circulation pattern 202
ocean current 204
oceanic cloud liquid water 198
oceanic integrated water vapour 198
oceanic surface wind speed 198
odour 29, 95
off-line method 9
olfactometer 95
olfactometry 95
ombrometer 64
on-line method 9
optical 43
optical depth 102
optical depth, aerosol 194
optical depth, total 194
overspeeding 58, 63, 174
ozone 76, 82, 102, 114, 159, 191
ozone profile 192

P

pan 121
particle 84, 89, 90, 91, 93, 94
particle, total suspended 85
particle imaging velocimetry 152
particle mass 85
particle size 88
particle sizer, optical 89, 90, 91
particle structure 94
passive instrument 16
passive measurement 17
passive method 157, 186
passive remote sensing 124
path-averaged 170, 197
path-averaging 129
pernix 43
peroxyacetyl 114
peroxy radical chemical amplifier 75
phosphor 78
photoacoustic 76
photoacoustic effect 81
photometer 76, 98, 99, 159, 191, 193
photometer, long path absorption 83
photometry/spectrometry 76
photovoltaic method 98
Piche evaporimeter 120
pilot balloon 20, 187, 188
piston manometer 52, 53
Pitot tube 59, 60
plane position indicator 171, 185
pluviograph 67
pluviometer 64
polar ice cover 198
polar stratospheric cloud (PSC) 187
positive temperature coefficient 34
potential evaporation 119, 121
Prandtl tube 59, 60
precipitation 29, 64, 69, 130, 132, 167, 184, 186, 187, 208
precipitation, global 198
precision 24
present weather sensor 65, 66, 68
pressure 29, 30, 47, 55, 59, 187
pressure balance 48, 52
pressure drop 85, 88
pressure plate 55, 57
pressure sensor 47
pressure sensor, electronic 48
pressure transducer 48
pressure transducer, capacitive 51
pressure transducer, electronic 53
pressure transducer, piezo-electric 52
pressure transducer, piezo-resistive 52
pressure transducer, quartz 52
pressure transducer, resistive 51
pressure tube 17, 59, 60
profiling, microwave 164

propeller 55
proton capture spectroscopy 92
proton transfer reaction 79
proton transfer reaction mass spectrometry 75
proton-induced x-ray emission 93
proton-transfer-reaction mass spectrometry 118
proxy data 4
psychrometer 42, 44, 46, 104
psychrometer, aspirated 17, 45
psychrometric difference 41
pyranometer 99, 100, 103, 104
pyrgeometer 99, 103
pyrheliometer 99, 100, 104
pyrometer 38
pyrradiometer 99, 103

Q
quality assurance 27
quality control 27
quantum cascade laser 75, 77, 156
quasi-photographic method 157

R
RADAR 6, 19, 126, 128, 129, 130, 137, 143, 149, 170, 172, 173, 181, 182, 184, 187, 188, 189
RADAR, rain 67, 68
RADAR, synthetic-aperture 132
radiation, infrared 37
radiation flux 98
radiation sensor 157
radiation shield 39, 46
radio-acoustic sounding system (RASS) 141
radioactive substance 96
radioactivity 29, 96, 97
radiometer 17, 30, 98, 102, 127, 157, 158, 159, 194, 195
radiometer, infrared 165
radiometer, interferometric 200
radiometer, microwave 164, 165, 167
radiometer, microwave scanning 158
radiometer, passive infrared 164
radiometer, space-borne 198
radiometer, spectral 98
radiometry 128
radiometry, satellite microwave 203

radiosonde 6, 18, 19, 46
rain 64, 69, 185
rain, type of 198
rain drop 65, 67, 143
rain gauge (totalisator) 64, 66
rain gauge, optical 69
rain gauge, ship 69
rain profile 186
rain RADAR, space-borne 186
rain sensor 64, 65
Raman LIDAR 143, 147, 148, 166, 168, 196
Raman polarization LIDAR 196
Raman scattering 127, 143, 147
range-resolving method 152
range detection 175
range determination 149, 150, 151
range height indicator 171, 186
range imaging 174
RASS 17, 128, 134, 135, 164, 177, 182
RASS, UHF 141
RASS, VHF 141
rawinsonde 20
Rayleigh backscatter 129, 148
Rayleigh scattering 127, 143, 159, 184
rain gauge, recording 64
radiation, reflected shortwave 101
refraction 131
refractivity 131
relaxed eddy accumulation (REA) method 117
remote sensing 1, 2, 4, 8, 124
remote sensing, active 124
representativity 23
resistance method 95
resolution 24
resonance fluorescence 75, 148
resonance method 85, 88, 95
response time 25

S
salinity 202, 203
saltation 94
saltiphon 94
sampler 85
satellite 1, 6, 18, 21, 22, 124, 133, 187, 192, 194, 201, 202
satellite, geostationary 22, 181
satellite, polar orbiting 22

Index

satellite, sun-synchronized 181
satellite image 157
saturation deficit 44
saturation water vapour pressure 41, 42, 45
scanning 8, 129
scanning electron microscope 94
scanning method 157
scanning mobility particle sizer 90
scan sounding 21
scatterometer 106, 171, 133, 199, 204
scintillation 97, 152
scintillation counter 97
scintillometer 152, 170, 180
scintillometer, crosswind 153
scintillometer, large aperture 152
scintillometer, small aperture 152
screen 38
sea ice 204
sea ice concentration 198
sea ice temperature 198
sea level rise 202
searchlight 187, 188
sea surface 133, 202
sediment concentration 204
selectivity 26
semiconductor 4
sensing 8
sensitivity 25, 31
sensor 10
sferic 205
shear wind 175
signal 10
signal-to noise-ratio 25, 131, 137, 144, 182
signal delay 149
silicon diaphragm 51
Six's thermometer 32, 33
slant wet delay 168
smoke detector, aspirated 86
SNODAR 140
snow 64, 65, 66, 69, 186, 201, 204
snow depth over sea ice 198
snow height 68
snow pillow 68
snow water equivalent 68, 198
snowflakes 186
SODAR 17, 126, 128, 135, 141, 143, 149, 170, 172, 173, 176, 177, 179, 180, 182
SODAR, bistatic 136

SODAR, mono-frequency 139, 172
SODAR, monostatic 136, 137
SODAR, multi-frequency 139
SODAR, phased-array 138, 139
SODAR-RASS 141
soil emission 108
soil heat flux 122
soil moisture 29, 64, 70, 72, 199
soil moisture, gravimetric 70
soil moisture ocean salinity mission 200
soil tension 71
solar irradiance, direct 100
solar radiation, direct 100
sonic 55
sonic thermometry 36
soot 86
sounding 8, 129
sounding, active 129
sounding, infrared 38
spaced antenna (SA) method 174
span 24, 31
specificity 26
spectral range, microwave 158
spectrometer 6, 76, 77, 83, 98, 127, 149, 154, 193
spectrometry, UV absorption 75
spectrometry, vacuum-ultraviolet (VUV) 75
spectrophotometer 102
spectroscope 77
spectroscopy, x-ray fluorescent 93
spherics 205
staring method 157
static chamber 109
Stevenson screen 18
Stokes scattering 148
subrefraction 132
sulphur 78
sulphur dioxide 114
sun photometer 102, 104, 159, 194
sunshine 104
sunshine autograph 6, 27
sunshine duration 104
sunshine recorder 99
superrefraction 131
surface chamber 107, 108
surface energy budget method 112
surface properties 198

A   Appendix: Technical guidelines and standards

Many detailed information on the usage of measurement instruments can be found in regularly updated guidelines and standards issued by national and international associations. The followings lists are by no means complete. They are intended to give an impression on the large variety of such information and can only be a starting point for further individual document research.

A.1   VDI and DIN (Germany)

(www.vdi.de)

VDI 2066
Part 1 (2006) Particulate matter measurement – Dust measurement in flowing gases – Gravimetric determination of dust load 2006-11.
Part 5 (1994) Particulate matter measurement – Dust measurement in flowing gases; particle size selective measurement by impaction method – Cascade impactor.
Part 8 (1995) Measurement of particles – Dust measurement in flowing gases – Measurement of smoke number in furnaces designed for EL-type fuel oil.
Part 10 (2004) Particulate matter measurement – Dust measurement in flowing gases – Measurement of PM_{10} and $PM_{2.5}$ emissions at stationary sources by impaction method.

VDI 2267
Part 1 (1999) Determination of suspended matters in ambient air – Measurement of As, Be, Cd, Co, Cr, Cu, Mn, Ni, Pb, Sb, Tl, Zn by atomic absorption spectrometry (AAS) after sampling on filters and digestion in an oxidizing acid mixture.
Part 2 (1983) Determination of suspended particulates in ambient air; measurement of lead by X-ray fluorescence.
Part 5 (1997) Determination of suspended matter in ambient air – Determination of the mass concentration of Be, Cd, Co, Cr, Cu, Fe, Mn, Ni, Pb, Sb, V, Zn by optical emission spectrometry (ICP-OES) after sampling on filters and digestion in an oxidising agent.
Part 7 (1988) Chemical analysis of particulates in ambient air; determination of thallium and its inorganic compounds as part of the dust precipitation by atomic absorption spectrometry.
Part 8 (2000) Determination of suspended particles in ambient air – Measurement of the mass concentration of mercury – Sampling by sorption as amalgam and determination by atomic absorption spectrometry (AAS) with cold vapour technique.
Part 9 (2002) Determination of suspended particulates matter in ambient air – Measurement of the mass concentration of mercury – Sampling by sorption as amalgam and determination by atomic fluorescense spectrometry (AFS) with cold vapour technique.

Part 12 (2008) Determination of suspended matter in ambient air – Measurement of As, Ca, Cd, Co, Cr, Cu, Fe, Mn, Ni, Pb, Sb and Zn by energy dispersive X-ray fluorescense (ED XRF).

Part 14 (2003) Determination of suspended matter in ambient air – Measurement of Al, As, Ca, Cd, Co, Cr, Cu, Fe, K, Mg, Mn, Na, Ni, Pb, V, Zn as part of dust deposition by optical emission spectrometry (ICP OES).

Part 15 (2005) Determination of suspended matter in ambient air – Measurement of the mass concentration of Al, As, Ca, Cd, Co, Cr, Cu, K, Mn, Ni, Pb, Sb, V, Zn as part of dust precipitation by mass spectrometry (ICP-MS).

Part 16 (2007) Determination of suspended matter in ambient air – Measurement of the mass concentration of As, Cd, Co, Cr, Cu, Ni, Pb, Sb, V und Zn as part of dust precipitation by atomic absorption spectrometry (AAS).

VDI 2268

Part 1 (1987) Chemical analysis of particulate matter; determination of Ba, Be, Cd, Co, Cr, Cu, Ni, Pb, Sr, V, Zn in particulate emissions by atomic spectrometric methods.

Part 2 (1990) Chemical analysis of particulate matter; determination of arsenic, antimony and selenium in dust emissions by atomic absorption spectrometry after separation of their volatile hydrides.

Part 3 (1988) Chemical analysis of particulate matter; determination of thallium in particulate emissions by atomic absorption spectrometry.

Part 4 (1990) Chemical analysis of particulate matter; determination of arsenic, antimony and selenium in dust emissions by graphite-furnace atomic absorption spectrometry.

VDI 2449

Part 1 (1995) Measurement methods test criteria – Determination of performance characteristics for the measurement of gaseous pollutants (immission).

Part 2 (1987) Basic concepts for characterization of a complete measuring procedure; glossary of terms.

Part 3 (2001) Measurement methods test criteria – General method for the determination of the uncertainty of calibratable measurement methods.

VDI 2451

Part 3 (1996) Measurement of gaseous immissions – Measurement of sulphur dioxide concentration – Tetrachloromercurate Pararosaniline (TCM) Method.

VDI 2452

Part 1 (1978) Air pollution measurement; measurement of total fluoride ion concentration; impinger method.

Part 2 (1975) Gaseous air pollution measurement; determination of fluoric ion concentration; preseparation and electrometric detection.

Part 3 (1987) Gaseous air pollution measurement; measurement of fluoride ion concentration; sorption method with prepared silver balls and heated membrane filter.

VDI 2453

Part 1 (1990) Gaseous air pollution measurement; determination of nitrogen dioxide concentration; photometric manual standard method (Saltzmann).

Part 2 (2002) Gaseous air pollution measurement – Measurement of concentration of nitrogen monoxide and nitrogen dioxide – Calibration of NO/NO_x chemiluminescence analysers using gas phase titration.

Part 3 (1995) Gaseous air pollution measurement – Determination of the nitrogen monoxide and nitrogen dioxide concentration – Preparation of the calibration gas mixtures and determination of their concentration.

VDI 2454

Part 1 (1982) Gaseous air pollution measurement; measurement of hydrogen sulphide concentration; molybdenum blue sorption method.

Part 2 (1982) Gaseous air pollution measurement; measurement of hydrogen sulphide concentration; methylene blue Impinger method.

VDI 2456

(2004) Stationary source emissions – Reference method for determination of the sum of nitrogen monoxide and nitrogen dioxide – Ion chromatography method.

VDI 2457

Part 1 (1997) Gaseous emission measurement – Chromatographic determination of organic compounds – Fundamentals.

Part 2 (1996) Gaseous emission measurement – Gas chromatographic determination of organic compounds – Sampling by absorption in a solvent (2-(2-methoxyethoxy)ethanol, methyldiglycol) at low temperature.

Part 3 (1996) Gaseous emission measurements – Gas chromatographic determination of organic compounds – Determination of substituted anilines – Sampling by solid phase adsorption.

Part 4 (2000) Gaseous emission measurement – Chromatographic determination of organic compounds – Sampling of acidic components in alkaline aqueous solution; Analysis by ion chromatography.

Part 5 (2000) Gaseous emission measurement – Chromatographic determination of organic compounds – Sampling in gas vessels, gas chromatographic analysis.

VDI 2459

Part 1 (2000) Gaseous emission measurement – Determination of carbon monoxide concentration using flame ionisation detection after reduction to methane.

VDI 2460

Part 1 (1996) Measurement of gaseous emissions – Infrared spectrometric determination of organic compounds – General principles.

Part 2 (1974) Gaseous emission measurement; infrared spectrometric determination of dimethyl formamide.

Part 3 (1981) Gaseous emission measurement; infrared spectrometric determination of cresol.

VDI 2461

Part 1 (1974) Gaseous air pollution measurement; measurement of ammonia gas concentration; indophenol method.

VDI 2462

Part 1 (1974) Gaseous emission measurement; determination of sulphur dioxide concentration; iodometric thiosulfate method.

Part 3 Gaseous emission measurement; determination of sulphur dioxide; gravimetric hydrogen peroxide method.

VDI 2463

Part 1 (1999) Particulate matter measurement – Gravimetric determination of mass concentration of suspended particulate matter in ambient air – General principles.

Part 4 (1976) Particulate matter measurement; Measurement of particulate matter in ambient air; LIB-filter method.

Part 7 (1982) Particulate matter measurement; measurement of mass concentration in ambient air; filter method; small filter device GS 050.

Part 8 (1982) Particulate matter measurement; measurement of mass concentration in ambient air; standard method for the comparsion of nonfractionating methods.

Part 11 (1996) Particulate matter measurement – Measurement of mass concentration in ambient air – Filter method; Digitel DHA-80 filter changer.

VDI 2465

Part 1 (1996) Measurement of soot (immission) – Chemical analysis of elemental carbon by extraction and thermal desorption of the organic carbon.

Part 2 (1999) Measurement of soot (Ambient Air) – Thermographical determination of elemental carbon after thermal desorption of organic carbon.

VDI 2466

Part 1 (2008) Gaseous emission measurement – Measurement of methane – Manual gas chromatography method.

Part 2 (2008) Gaseous emission measurement – Measurement of methane – Automatic method – Flame ionisation detector (FID).

VDI 2467

Part 2 (1991) Gaseous air pollution measurement; measurement of primary and secondary aliphatic amines by means of the high performance liquid chromatography (HPLC).

VDI 2468

Part 7 (2006) Ambient air measurement – Measurement of peroxyacetyl nitrate (PAN) – Gas chromatography method.

Part 8 (2006) Ambient air measurement – Measurement of peroxyacetyl nitrate (PAN) – Preparation of PAN calibration gas.

Part 9 (1995) Gaseous air pollution measurement – Measurement of hydrogen peroxide – Continuous fluorometric method.

Part 10 (1995) Gaseous air pollution measurement – Preparation of hydrogen peroxide calibration – Gas mixtures.

VDI 2469

Part 1 (2005) Gaseous emission measurement – Measurement of nitrous oxide – Manual gas chromatographic method.

Part 2 (2005) Gaseous emission measurement – Measurement of nitrous oxide – Automatic infrared spectrometric method.

VDI 2470
Part 1 (1975) Gaseous emission measurement; measurement of gaseous fluorine compounds; absorption method.

VDI 3481
Part 2 (1998) Gaseous emission measurement – Determination of gaseous organic carbon in waste gases – Adsorption on silica gel.
Part 3 (1995) Gaseous emission measurement – Determination of volatile organic compounds, especially solvents, flame ionization detector (FID).
Part 4 (2007) Gaseous emission measurement – Measurement of the concentrations of total organic carbon and methane carbon using the flame ionisation detector (FID).

VDI 3483
Part 1 (1979) Gaseous air pollution measurement; determination of total organic compounds by use of a flame ionization detector (FID); fundamentals.

VDI 3485
Part 1 (1988) Ambient air measurement; measurement of gaseous phenoloc compounds; p-nitroaniline method.

VDI 3486
Part 1 (1979) Measurement of gaseous emissions; Measurement of the hydrogen sulfide concentration; potentiometric titration method.
Part 2 (1979) Measurement of gaseous emission; Measurement of the hydrogen sulfide concentration; iodometric titration method.

VDI 3487
Part 1 (1978) Gaseous emission measurement; measurement of carbon disulfide concentration; iodometric titration method.

VDI 3488
Part 1 (1979) Gaseous emission measurement; measurement of chlorine and oxides of chlorine; methyl orange method.
Part 2 (1980) Gaseous emission measurement; measurement of chlorine concentration; bromide iodide method.

VDI 3492
(2004) Indoor air measurement – Ambient air measurement – Measurement of inorganic fibrous particles – Scanning electron microscopy method.

VDI 3493
Part 1 (1982) Gaseous emission measurement; determination of vinyl chloride; gas chromatographic method; grab sampling.

VDI 3496
Part 1 (1982) Gaseous emission measurement; determination of basic nitrogen compounds seizable by absorption in sulphuric acid.

VDI 3497

Part 3 (1988) Determination of particulate anions in ambient air; analysis of chloride, nitrate, and sulfate by ion chromatography using suppressor technique after aerosol sampling on PTFE-filters.

Part 4 (1991) Determination of particulate anions in ambient air; analysis of chloride, nitrate, and sulphate by ion chromatography using single-column-technique after aerosol sampling on PTFE-filters.

VDI 3498

Part 1 (2002) Ambient air measurement – Indoor air measurement – Measurement of polychlorinated dibenzo-p-dioxins and dibenzofurans; Method using large filters.

Part 2 (2002) Ambient air measurement – Indoor air measurement – Measurement of polychlorinated dibenzo-p-dioxins and dibenzofurans; Method using small filters.

VDI 3783

Part 8 (2002) Environmental meteorology – Turbulence parameters for dispersion models supported by measurement data.

VDI 3786

Part 1 (1995) Environmental meteorology – Meteorological measurements – Fundamentals.

Part 2 (2000) Environmental meteorology – Meteorological measurements concerning questions of air pollution – Wind.

Part 3 (1985) Meteorological measurements concerning questions of air pollution; air temperature.

Part 4 (1985) Meteorological measurements concerning questions of air pollution; air humidity.

Part 5 (1986) Meteorological measurements concerning questions of air pollution; global radiation, direct solar radiation and net total radiation.

Part 6 (1983) Meteorological measurements of air pollution; turbidity of ground-level atmosphere standard visibility.

Part 7 (2009) Environmental meteorology – Meteorological measurements – Precipitation.

Part 8 (1987) Meteorological measurements; concerning questions of air pollution; aerological measurements.

Part 9 (2007) Environmental Meteorology – Meteorological measurements – Visual weather observations.

Part 10 (1994) Environmental meteorology; measurement of the atmospheric turbidity due to aerosol particles with sunphotometers.

Part 11 (1994) Environmental meteorology; determination of the vertical wind profile by Doppler SODAR systems.

Part 12 (2008) Environmental meteorology – Meteorological measurements – Turbulence measurements with sonic anemometers.

Part 13 (2006) Environmental meteorology – Meteorological measurements – Measuring station.

Part 14 (2001) Environmental meteorology – Ground-based remote sensing of the wind vector – Doppler wind LIDAR.

Part 15 (2004) Environmental meteorology – Ground-based remote sensing of visual range – Visual-range LIDAR.

Part 16 (2009) Environmental meteorology – Meteorological measurements – Atmospheric pressure.

Part 17 (2007) Environmental meteorology – Ground-based remote sensing of the wind vector – Wind profiler radar.
Part 18 (2009) Environmental meteorology – Ground-based remote sensing of of temperature – Radio-acoustic sounding systems (RASS).

VDI 3789
Part 2 (1994) Environmental meteorology – Interactions between atmosphere and surfaces – Calculation of short-wave and long-wave radiation.
Part 3 (2001) Environmental meteorology – Interactions between atmosphere and surfaces – Calculation of spectral irradiances in the solar wavelength range.

VDI 3790
Part 1 (2005) Environmental meteorology – Emissions of gases, odours and dusts form diffuse sources – Fundamentals.
Part 2 (2000) Environmental meteorology – Emissions of gases, odours and dusts from diffuse sources – Landfills.
Part 3 (2008) Environmental meteorology – Emission of gases, odours and dusts from diffuse sources – Storage, transshipment and transportation of bulk materials.

VDI 3861
Part 1 (1989) Measurement of fibrous particles; manual measurement of asbestos in flowing clean exhaust gas; determination of asbestos mass concentration by IR-spectroscopy.
Part 2 (2008) Stationary source emissions – Measurement of inorganic fibrous particles in exhaust gas – Scanning electron microscopy method.

VDI 3869
Part 3 (2008) Measurement of ammonia in ambient air – Sampling with diffusion separators (denuders) – Photometric or ion chromatographic analysis.
Part 4 (1996) Measurement of acids and bases in ambient air – Measurement of ammonia; sampling in diffusion separators coated with citric acid – Determination by indophenol method.

VDI 3870
Part 10 (1994) Analysis of rainwater – Determination of the pH value of rainwater.
Part 11 (1996) Analysis of rainwater – Determination of free acidity.
Part 13 (1996) Analysis of rainwater – Determination of chloride, nitrate and sulfate in rainwater by ion chromatography using the suppressor technique.

VDI 3882
Part 1 (1992) Olfactometry; determination of odour intensity.
Part 2 (1994) Olfactometry – Determination of hedonic odour tone.

VDI 3940
Part 1 (2006) Measurement of odour impact by field inspection – Measurement of the impact frequency of recognizable odours – Grid measurement.
Part 2 (2006) Measurement of odour impact by field inspection – Measurement of the impact frequency of recognizable odours – Plume measurement.
Part 3 (2008) Measurement of odour in ambient air by field inspections – Determination of odour intensity and hedonic odour tone.

VDI 4210

Part 1 (1999) Remote sensing – Atmospheric measurements with LIDAR – Measuring gaseous air pollution with DAS LIDAR.

VDI 4211

Part 1 (2000) Remote sensing – Atmospheric measurements near ground wit FTIR spectroscopy – Measurement of gaseous emissions and immissions; Fundamentals.

VDI 4212

Part 1 (2008) Remote sensing – Atmospheric measurements near ground with DOAS – Gaseous emissions and ambient air measurements – Fundamentals.

VDI 4251

Part 1 (2007) Measurement of airborne microorganisms and viruses in ambient air – Plume measurement.

VDI 4252

Part 2 (2004) Measurement of airborne microorganisms and viruses in ambient air – Active sampling of bioaerosols – Separation of airborne mould on gelatine/polycarbonate filters.

Part 3 (2008) Measurement of airborne microorganisms and viruses in ambient air – Active sampling of bioaerosols – Separation of airborne bacteria with impingers using the principle of critical nozzle.

VDI 4253

Part 2 (2004) Measurement of airborne microorganisms and viruses in ambient air – Culture based method for the determination of the concentration of mould in air – Indirect method after sampling with gelatine/polycarbonate filters.

Part 3 (2008) Measurement of airborne microorganisms and viruses in ambient air – Culture based method for the quantitative determination of bacteria in air – Method after separation in liquids.

VDI 4280

Part 1 (1996) Planning of ambient air quality measurements – General rules.

Part 2 (2000) Planning of ambient air quality measurements – Rules for planning investigations of traffic related air pollutants in key pollution areas.

Part 3 (2003) Planning of ambient air quality measurements – Measurement strategies for the determination of air quality characteristics in the vicinity of stationary emission sources.

Part 4 (2009) Planning of ambient air quality measurements – Substitution of missing values in series of measured values of ambient air quality.

Part 5 Planning of ambient air quality measurements – Evaluation of the uncertainty of spatial air quality assessments.

VDI 4285

Part 1 (2005) Determination of diffusive emissions by measurement – Basic concepts.

Part 2 (2006) Determination of diffusive emissions by measurements – Industrial halls and livestock farming.

VDI 4320

Part 1 (2008) Measurement of atmospheric depositions – Sampling with Bulk- and Wet-Only-collectors – General principles.

DIN 33962

(1997) Measurement of gaseous emissions – Automatic measurement systems for single measurements of nitrogen monooxide and nitrogen dioxide.

Updated overviews on these VDI guidelines (many of them are available in English) and German (DIN) standards may be obtained from: http://www.vdi.de/7636.0.html?&L=1.

A.2 Other German associations

DVWK Deutscher Verband für Wasserwirtschaft und Kulturbau e.V., Bonn (http://www.dwa.de)

DVWK-Merkblatt 238/1996 Ermittlung der Verdunstung von Land- und Wasserflächen (Determination of evaporation from land and lake surfaces)

KTA Kerntechnischer Ausschuss (die Geschäftsstelle ist beim Bundesamt für Strahlenschutz, Salzgitter) (http://www.kta-gs.de)

KTA 1503 Überwachung der Ableitung radioaktiver Stoffe (Surveillance of radioactive substances)

KTA 1508 Instrumentierung zur Ermittlung der Ausbreitung radioaktiver Stoffe in der Atmosphäre (Instruments for the determination of dispersion parameters for radioactive substances in the atmosphere)

A.3 British Standards Institution BSI

(www.bsigroup.com)

BS 1339-3:2004. Humidity. Guide to the measurement of humidity

BS 7440:1991, ISO 9059:1990. Method for calibrating field pyrheliometers by comparison to a reference pyrheliometer

BS 7621:1993, ISO 9847:1992. Method for calibrating field pyranometers by comparison to a reference pyranometer

BS 7843-1.1:1996. Guide to the acquisition and management of meteorological precipitation data. Network design. The user requirement for precipitation data

BS 7843-1.2:1996. Guide to the acquisition and management of meteorological precipitation data. Network design. Network design and monitoring

BS 7843-2.1:1996. Guide to the acquisition and management of meteorological precipitation data. Field practices and data management. Technical aspects in the field

BS 7843-2.2:1996. Guide to the acquisition and management of meteorological precipitation data. Field practices and data management. Methods of observation and data tabulation

BS 7843-2.3:1996. Guide to the acquisition and management of meteorological precipitation data. Field practices and data management. Data management

BS 7843-2.4:1996. Guide to the acquisition and management of meteorological precipitation data. Field practices and data management. Areal rainfall

BS 7843-3.1:1999. Guide to the acquisition and management of meteorological precipitation data. Specification for raingauges. Storage raingauges. Storage raingauges

BS 7843-3.2:2005. Guide to the acquisition and management of meteorological precipitation data. Specification for raingauges. Tipping bucket gauges

A.4   American National Standards Institute ANSI

(www.ansi.org)

ANSI Z136 series: laser eye safety

ASTM D3631-99(2007) Standard Test Methods for Measuring Surface Atmospheric Pressure

ASTM D3796-90(2004) Standard Practice for Calibration of Type S Pitot Tubes

ASTM D5096-02(2007) Standard Test Method for Determining the Performance of a Cup Anemometer or Propeller Anemometer

ASTM D5527-00(2007) Standard Practices for Measuring Surface Wind and Temperature by Acoustic Means (sonic)

ASTM D5741-96(2007)e1 Standard Practice for Characterizing Surface Wind Using a Wind Vane and Rotating Anemometer

ASTM D7145-05 Standard Guide for Measurement of Atmospheric Wind and Turbulence Profiles by Acoustic Means (SODAR)

ASTM E337-02(2007) Standard Test Method for Measuring Humidity with a Psychrometer (the Measurement of Wet- and Dry-Bulb Temperatures)

A.5   European Committee for Standardization (CEN)

(http://www.cen.eu/cenorm/homepage.htm)

TC264

EN 12341:1998 Air quality – Determination of the PM 10 fraction of suspended particulate matter – Reference method and field test procedure to demonstrate reference equivalence of measurement methods

EN 13528-1:2002 Ambient air quality – Diffusive samplers for the determination of concentrations of gases and vapours – Requirements and test methods – Part 1: General requirements

EN 13528-2:2002 Ambient air quality – Diffusive samplers for the determination of concentrations of gases and vapours – Requirements and test methods – Part 2: Specific requirements and test methods

EN 13528-3:2003 Ambient air quality – Diffusive samplers for the determination of concentrations of gases and vapours – Requirements and test methods – Part 3: Guide to selection, use and maintenance

EN 13725:2003/AC:2006 Air quality – Determination of odour concentration by dynamic olfactometry

EN 14211:2005 Ambient air quality – Standard method for the measurement of the concentration of nitrogen dioxide and nitrogen monoxide by chemiluminescence

EN 14212:2005 Ambient air quality – Standard method for the measurement of the concentration of sulphur dioxide by ultraviolet fluorescence

EN 14625:2005 Ambient air quality – Standard method for the measurement of the concentration of ozone by ultraviolet photometry

EN 14626:2005 Ambient air quality – Standard method for the measurement of the concentration of carbon monoxide by nondispersive infrared spectroscopy

EN 14662-1:2005 Ambient air quality – Standard method for measurement of benzene concentrations – Part 1 : Pumped sampling followed by thermal desorption and gas chromatography

EN 14662-2:2005 Ambient air quality – Standard method for measurement of benzene concentrations – Part 2 : Pumped sampling followed by solvent desorption and gas chromatography

EN 14662-3:2005 Ambient Air Quality – Standard method for the measurement of benzene concentrations – Part 3: Automated pumped sampling with in situ gas chromatography

EN 14662-4:2005 Ambient air quality – Standard method for measurement of benzene concentrations – Part 4: Diffusive sampling followed by thermal desorption and gas chromatography

EN 14662-5:2005 Ambient air quality – Standard method for measurement of benzene concentrations – Part 5: Diffusive sampling followed by solvent desorption and gas chromatography

EN 14902:2005/AC:2006 Ambient air quality – Standard method for the measurement of Pb, Cd, As and Ni in the PM10 fraction of suspended particulate matter

EN 14907:2005 Ambient air quality – Standard gravimetric measurement method for the determination of the PM2,5 mass fraction of suspended particulate matter

EN 15483:2008 Ambient air quality – Atmospheric measurements near ground with FTIR spectroscopy

EN 15549:2008 Air quality – Standard method for the measurement of the concentration of benzo[a]pyrene in ambient air

A.6 International Organisation for Standardization (ISO)

(http://www.iso.org/iso/iso_catalogue/catalogue_tc.htm)

TC146/SC2
ISO 7708:1995 Air quality – Particle size fraction definitions for health-related sampling
TC146/SC3
ISO 4219:1979 Air quality – Determination of gaseous sulphur compounds in ambient air – Sampling equipment
ISO 4220:1983 Ambient air – Determination of a gaseous acid air pollution index – Titrimetric method with indicator or potentiometric end-point detection

ISO 4221:1980 Air quality – Determination of mass concentration of sulphur dioxide in ambient air – Thorin spectrophotometric method
ISO 4224:2000 Ambient air – Determination of carbon monoxide – Non-dispersive infrared spectrometric method
ISO 6767:1990 Ambient air – Determination of the mass concentration of sulfur dioxide – Tetrachloromercurate (TCM)/pararosaniline method
ISO 6768:1998 Ambient air – Determination of mass concentration of nitrogen dioxide – Modified Griess-Saltzman method
ISO 7996:1985 Ambient air – Determination of the mass concentration of nitrogen oxides – Chemiluminescence method
ISO 8186:1989 Ambient air – Determination of the mass concentration of carbon monoxide – Gas chromatographic method
ISO 9835:1993 Ambient air – Determination of a black smoke index
ISO 9855:1993 Ambient air – Determination of the particulate lead content of aerosols collected on filters – Atomic absorption spectrometric method
ISO 10312:1995 Ambient air – Determination of asbestos fibres – Direct transfer transmission electron microscopy method
ISO 10313:1993 Ambient air – Determination of the mass concentration of ozone – Chemiluminescence method
ISO 10473:2000 Ambient air – Measurement of the mass of particulate matter on a filter medium – Beta-ray absorption method
ISO 10498:2004 Ambient air – Determination of sulfur dioxide – Ultraviolet fluorescence method
ISO 12884:2000 Ambient air – Determination of total (gas and particle-phase) polycyclic aromatic hydrocarbons – Collection on sorbent-backed filters with gas chromatographic/mass spectrometric analyses
ISO 13794:1999 Ambient air – Determination of asbestos fibres – Indirect-transfer transmission electron microscopy method
ISO 13964:1998 Air quality – Determination of ozone in ambient air – Ultraviolet photometric method
ISO 14965:2000 Air quality – Determination of total non-methane organic compounds – Cryogenic preconcentration and direct flame ionization detection method
ISO 14966:2002 Ambient air – Determination of numerical concentration of inorganic fibrous particles – Scanning electron microscopy method (Cor 1:2007)
ISO 15337:2009 Ambient air – Gas phase titration – Calibration of analysers for ozone
ISO 16362:2005 Ambient air – Determination of particle-phase polycyclic aromatic hydrocarbons by high performance liquid chromatography

TC146/SC4
ISO 4225:1994 Air quality – General aspects – Vocabulary
ISO 4226:2007 Air quality – General aspects – Units of measurement
ISO 7168-1:1999 Air quality – Exchange of data – Part 1: General data format
ISO 7168-2:1999 Air quality – Exchange of data – Part 2: Condensed data format
ISO 8756:1994 Air quality – Handling of temperature, pressure and humidity data
ISO 9169:2006 Air quality – Definition and determination of performance characteristics of an automatic measuring system
ISO 9359:1989 Air quality – Stratified sampling method for assessment of ambient air quality

ISO 11222:2002 Air quality – Determination of the uncertainty of the time average of air quality measurements
ISO/DIS 11771 Air quality – Determination of time-averaged mass emissions and emission factors – General approach
ISO 13752:1998 Air quality – Assessment of uncertainty of a measurement method under field conditions using a second method as reference
ISO 14956:2002 Air quality – Evaluation of the suitability of a measurement procedure by comparison with a required measurement uncertainty
ISO 20988:2007 Air quality – Guidelines for estimating measurement uncertainty

TC146/SC5
ISO 16622 on the use of sonic anemometers/thermometers for mean wind measurements
ISO 17713-1 on wind tunnel test methods for rotating anemometer performance
ISO 17713-2 wind tunnel test methods for wind vanes
ISO 17714 test methods for comparing the performance of thermometer shields/screens.

TC180/SC1
ISO 9059:1990 Solar energy – Calibration of field pyrheliometers by comparison to a reference Pyrheliometer
ISO 9060:1990 Solar energy – Specification and classification of instruments for measuring hemispherical solar and direct solar radiation
ISO 9845-1:1992 Solar energy – Reference solar spectral irradiance at the ground at different receiving conditions – Part 1: Direct normal and hemispherical solar irradiance for air mass
ISO 9846:1993 Solar energy – Calibration of a pyranometer using a Pyrheliometer
ISO 9847:1992 Solar energy – Calibration of field pyranometers by comparison to a reference Pyranometer
ISO/TR 9901:1990 Solar energy – Field pyranometers – Recommended practice for use

A.7   International Electrotechnical Commision IEC

(www.iec.ch)

IEC 60825-1 Safety of laser products – Part 1: Equipment classification, requirements and user's guide.

A.8   Weather Services and WMO

Many Weather Services provide documentations for the operation of synoptic and upper air stations. Make requests to your national weather service.

Manuals of the World Meteorological Organization in Geneva
(http://www.wmo.ch/pages/themes/wmoprod/manuals.html)
306–Manual on codes – International codes
Volume I.1 Part A: Alphanumeric codes
Volume I.2 Part B: Binary codes and Part C: Common features to binary and alpha-numeric codes
Volume II – Regional codes and national coding practices
332–Manual for estimation of probable maximum precipitation
386–Manual on the Global Telecommunication System
Volume I – Global aspects Volume II – Regional aspects
407–International cloud atlas
Volume I – Manual on the observation of clouds and other meteors
Volume II (plates)
485–Manual on the Global Data-processing System
Volume I – Global aspects, 1991
Volume II – Regional aspects, 1992
519–Manual on stream gauging
Volume I – Fieldwork; xiv + 308 pp.
Volume II – Computation of discharge; xiv + 258 pp.
544–Manual on the Global Observing System
Volume I – Global aspects
Volume II – Regional aspects
558–Manual on marine meteorological services
Volume I – Global aspects
Volume II – Regional aspects
948–Manual on Sediment Management and Measurement

Index to the Appendix

A
absorption in sulphuric acid 245
absorption method 245
acids 247
adsorption on silica gel 245
aerological measurements 246
airborne microorganisms 248
air humidity 246
air pollution 242, 248
air pollution index 251
air quality 248, 250, 252, 253
air temperature 246
aliphatic amines 244
ammonia 243, 247
areal rainfall 250
asbestos fibres 252
atomic absorption spectrometric method 252
atomic absorption spectrometry 241, 242
atomic fluorescense spectrometry 241
atomic spectrometric methods 242

B
bases 247
benzene 251
benzo[a]pyrene 251
beta-ray absorption method 252
black smoke index 252
bromide iodide method 245
bulk-collectors 249

C
carbon disulfide 245
carbon monoxide 243, 251, 252
cascade impactor 241
chemiluminescence 243, 251
chemiluminescence method 252
chloride 246, 247
chlorine 245
chromatographic determination 243
computation of discharge 254
cup anemometer 250

D
DAS LIDAR 248
deposition 249
diffuse source 247
diffusive emissions 248
diffusive samplers 250, 251
direct solar radiation 246, 253
dispersion 249
DOAS 248
Doppler wind LIDAR 246
dry-bulb temperatures 250
dust precipitation 242
dusts 247

E
electron microscopy method 252
emission factors 253
energy dispersive X-ray fluorescense 242
evaporation 249

F
fibrous particles 247
filter method 244
filters 246
flame ionisation detection 243, 244, 245
flame ionization detection method 252
flame ionization detector 245
fluoric ion concentration 242
fluoride ion concentration 242
fluorine compounds 245
fluorometric method 244
free acidity 247
FTIR spectroscopy 248, 251

G
gas chromatographic analysis 243
gas chromatographic/mass spectrometric analyses 252
gas chromatographic method 244, 245, 252
gas chromatography 244, 251
gaseous emissions 243, 249
gaseous immissions 242
gas phase titration 252
global data-processing system 254
global observing system 254
global radiation 246
global telecommunication system 254
graphite-furnace atomic absorption spectrometry 242
gravimetric determination 241, 244
gravimetric hydrogen peroxide method 244
gravimetric measurement 251
grid measurement 247

Griess-Saltzman method 252

H
hedonic odour tone 247
high performance liquid chromatography 244, 252
humidity 249, 250, 252
hydrogen peroxide 244
hydrogen sulfide 243, 245

I
impaction method 241
impinger method 242
indoor air measurement 246
indophenol method 243, 247
infrared spectrometric determination 243
infrared spectrometric method 244
inorganic fibrous particles 245, 252
international cloud atlas 254
iodometric thiosulfate method 244
iodometric titration method 245
ion chromatographic analysis 247
ion chromatography 243, 246, 247
ion chromatography method 243
IR-spectroscopy 247

L
laser eye safety 250
LIB-filter method 244
LIDAR 248
long-wave radiation 247

M
marine meteorological services 254
mass emissions 253
mass spectrometry 242
measuring station 246
methane 244
methane carbon 245
methylene blue impinger method 243
methyl orange method 245
microorganisms 248
missing values 248
molybdenum blue sorption method 243

N
net total radiation 246
nitrate 246, 247
nitrogen 245

nitrogen dioxide 242, 243, 249, 251, 252
nitrogen monooxide 243, 249, 251
nitrogen oxides 252
nitrous oxide 244
non-dispersive infrared spectrometric method 252
non-methane organic compounds 252
nondispersive infrared spectroscopy 251

O
odour 247, 251
odour impact 247
odour intensity 247
olfactometry 247, 251
optical emission spectrometry 241, 242
organic carbon 245
ozone 251, 252

P
p-nitroaniline method 245
particles 241
particle size fraction definitions 251
particulate lead 252
particulate matter 241, 242, 244, 252
particulates 241
performance characteristics 242
peroxyacetyl nitrate 244
phenoloc compounds 245
photometric manual standard method 242
pH value 247
Pitot Tubes 250
plume measurement 247, 248
PM10 250, 251
PM2,5 251
polycyclic aromatic hydrocarbons 252
potentiometric titration method 245
precipitation 246, 249, 250, 254
pressure 246, 250, 252
propeller anemometer 250
psychrometer 250, 253
pyranometers 249, 253
pyrheliometer 249, 253

R
radio-acoustic sounding systems (RASS) 247
radioactive substances 249
rainwater 247
rotating anemometer 250, 253

S

safety of laser 253
scanning electron microscopy 247
scanning electron microscopy method 245, 252
sediment management 254
short-wave radiation 247
smoke number 241
SODAR 246, 250
solar energy 253
solar irradiance 253
solar spectral irradiance 253
solid phase adsorption 243
solvent desorption 251
sonic 250
sonic anemometers 246, 253
soot 244
spectral irradiances 247
stationary emission sources 248
storage raingauges 250
stream gauging 254
sulfate 246, 247
sulfur dioxide 252
sulphate 246
sulphur 251
sulphur dioxide 242, 244, 251, 252
sun photometers 246
suspended matter 241, 242
suspended particles 241
suspended particulates 241

T

temperature 247, 250, 252
tetrachloromercurate (TCM)/pararosaniline method 252
tetrachloromercurate pararosaniline (TCM) method 242
thermal desorption 244, 251
thermographical determination 244
thermometers 253
thermometer shields/screens 253
thorin spectrophotometric method 252
tipping bucket gauges 250
titrimetric method 251
total organic compounds 245
turbidity 246
turbulence 246, 250

U

ultraviolet fluorescence 251
ultraviolet fluorescence method 252
ultraviolet photometric method 252
ultraviolet photometry 251
uncertainty 248, 253
uncertainty of calibratable measurement methods 242

V

vertical wind profile 246
vinyl chloride 245
viruses 248
visibility 246
visual-range LIDAR 246
visual range 246
visual weather observations 246
volatile organic compounds 245

W

wet-only-collectors 249
wet-temperatures 250
wind 246, 250, 253
wind profiler radar 247
wind tunnel 253
wind vane 250, 253
wind vector 246, 247

X

X-ray fluorescence 241

SLS Series Scintillometers

Heat and Momentum Flux by Purely Optical Means

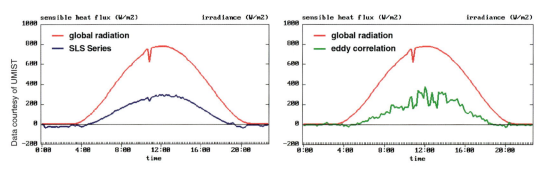

The SLS Series Scintillometers measure turbulence, heat flux and momentum flux by purely optical means. In combination with other meteorological sensors the system can determine latent heat flux or evaporation.

- virtually no statistical noise
- true path averaging
- no external wind sensor needed
- no roughness length estimate needed
- stable, unstable, near neutral conditions

Scintec

www.scintec.com - info@scintec.com